Manual of Water Supply Practices

M77

Condition Assessment of Water Mains

American Water Works Association

Manual of Water Supply Practices—M77

Condition Assessment of Water Mains

If you find errors in this manual, please email books@awwa.org. Possible errata will be posted at www.awwa.org/resources-tools/resource-development-groups/manuals-program.aspx.

Managing Editor – Book Products: Melissa Valentine
Project Manager/Technical Editor: Suzanne Snyder
Cover Art: Michael Labruyere
Production: Innodata
Senior Specialist – Manuals: Willadee Hitchcock

Library of Congress Cataloging-in-Publication Data
Names: Ellison, Dan, author. | American Water Works Association, issuing body.
Title: Condition assessment of water mains / by Dan Ellison.
Other titles: AWWA manual ; M77.
Description: First edition. | Denver, CO : American Water Works Association, [2019] | Series: Manual of water supply practices ; M77 | Includes bibliographical references and index.
Identifiers: LCCN 2019001636 | ISBN 9781625763310 (alk. paper)
Subjects: LCSH: Water-pipes--Monitoring.
Classification: LCC TD491 .E42 2019 | DDC 628.1/5072--dc23
LC record available at https://lccn.loc.gov/2019001636

Printed in the United States of America

ISBN 978-1-62576-331-0 eISBN-13 978-1-61300-503-3

American Water Works Association

American Water Works Association
6666 West Quincy Avenue
Denver, CO 80235-3098
awwa.org

Contents

List of Figures, vii

List of Tables, xi

Preface, xiii

Acknowledgments, xv

Chapter 1 The Benefits of Condition Assessment for Water Mains. 1

Definition of Condition Assessment, 2
The Benefits and Costs of Condition Assessment, 3
Condition Assessment as Part of an Assess-and-Fix Strategy, 6
Additional Considerations When Using Condition Assessment, 7
References, 8

Chapter 2 Building Support for a Condition Assessment Program. 9

Making a Business Case for Condition Assessment, 9
Accounting Treatment: Ways to Account for Condition Assessment Costs, 12
Budgeting for a Systematic Condition Assessment Program, 13
Gaining Support From Policy Makers, The Public, and Other Stakeholders, 14
Summary, 16
References, 16

Chapter 3 Planning a Condition Assessment Project or Program 19

Levels of Condition Assessment, 19
Choosing an Assessment Method, 21
Available Technologies, 23
Planning of Field Work/Communications, 27
Contract Guidelines, 29
Summary, 30
Reference, 30

Chapter 4 Desktop Condition Assessment. 31

Overview, 32
Data Cleanup and Assessment, 32
Setting Level of Service Goals, 36
Evaluating the Likelihood of Failure Desktop Score, 37
Long-Term Renewal Planning, 43

Chapter 5 External Corrosion Surveys. 47

External Corrosion Direct Assessment, 47
Economic Benefits, 55
References, 55

Chapter 6 Spot Assessment 57
Strategies for Effective Spot Assessment, 58
Testing Procedures, 60
Post Assessment and Interpretation of Results, 73
References, 75

Chapter 7 Leak Detection 77
Why Leak Data Are Important for Condition Assessment, 78
How and When to Perform Leak Detection, 79
Acoustic Leak Detection, 79
Non-acoustic Leak Detection, 91
Summary, 96
References, 96

Chapter 8 Internal Remote Visual Inspection 97
Internal Visual Inspection as a Condition Assessment Tool, 98
Recent Advances in Remote Visual Inspection Technology, 99
Logistical Considerations for Performing Remote Internal Visual Inspection, 100
Current Methods and Technologies Available for Internal Visual Inspection, 101
Implementation Strategies for Internal Remote Visual Inspection, 110
Summary, 112

Chapter 9 Physical Entry Inspections 115
Overview: Physical Entry Inspections, 116
Summary, 124
References, 124

Chapter 10 Acoustic Velocity Testing 125
History, 126
Background, 127
Performance, 130
Requirements and Limitations, 132
References, 133

Chapter 11 Electromagnetic Testing Technologies 135
History and Current Use of EM Testing in Water Mains, 136
Technology Overview, 140
Defect Resolution Considerations, 146
Advantages and Limitations, 146
References, 147

Chapter 12 Magnetic Flux Leakage 149
Current State of the Art, 150
MFL Application Considerations and Inspection Logistics, 150
Data Output and Post Processing, 157
Planning and Delivering a Quality MFL Project, 160
Reference, 162

Chapter 13 Condition Assessment of PCCP 163
PCCP Construction, 163
PCCP Deterioration And Failure Modes, 165
Inspection and Assessment, 166

Strategies for Managing Distressed PCCP, 178
References, 182

Chapter 14 Hydrostatic (Pressure) Testing of Existing Pipes 185
The Basics, 186
Hydrostatic Examination, 187
Procedure Development, 188
Pressure Limits, 190
After the Initial Tests, 191
References, 191

Chapter 15 Strategies for Economical Assessments of Low-Value Pipes 193
Nondisruptive Sampling Methods, 194
Nondisruptive Screening Techniques, 195
Leveraging Condition Assessment Data, 196
The Assess-and-Fix Approach, 197
Forensic Analysis of Breaks, 198
Reference, 200

Chapter 16 The Next Steps: Using Condition Assessment Information..........201
Purpose of Condition Assessment Information, 201
Leveraging Data, 202
Data Analysis, 203
Presentation of Results, 203
Example Approaches, 204
References, 206

Appendix A Other Assessment Methods 207
Acoustic Resonance Testing, 207
Pipe Wall Analysis Using Magnetometry, 207

Index 209

List of AWWA Manuals 215

This page intentionally blank.

Figures

2-1 Aerial view of Anchorage, Alaska, railroad yard, 11

2-2 Progressive condition assessment and resulting actions, 14

3-1 The tradeoffs between costs, coverage, and degree of inspection, 20

4-1 Decision tree process for GIS CoF scoring, 40

4-2 Risk matrix, 41

4-3 Example of color-coded risk mapping, 42

4-4 Example decay curve based on two known points, 43

4-5 Example of optimizing pipe EUL by considering CoF scores, 44

4-6 Example of using GIS to predict funding needs based on service levels, 45

5-1 Graphical representation of soil resistivity fluctuations along a pipeline alignment as measured using Emag and comparison to Wenner 4-pin results, 50

5-2 Graphical representation of cell-to-cell, close-interval survey results along pipeline alignment, 52

5-3 Depth of cover, soil resistivity, and close interval potential survey indirect inspection data aligned along pipeline rights-of-way, 54

6-1 Measuring the depth of corrosion pits in cast-iron pipe, 62

6-2 In situ pH testing of concrete pipe using phenolphthalein, 62

6-3 Ultrasonic thickness measurement, 64

6-4 Potential mapping, 65

6-5 In situ electrochemical testing, 66

6-6 Sampling slime for MIC testing, 67

6-7 Identifying corrosion products using SEM/EDS, 69

6-8 Optical microscopic pitted surface (top); pit depth profiling (bottom), 70

6-9 Comparative metallographic analysis of heat-treated (welded) zones, 71

6-10 Petrographic analysis of concrete, illustrating cracks, 71

6-11 ASTM B117 test on heat-treated areas of stainless steel (1000 h of testing in salt chamber), 72

6-12 Phenolphthalein stain testing of asbestos cement pipe, 73

7-1 Leak detection techniques covered in this chapter, 80

7-2 A leak detection monitor installed in a below-ground chamber, 84

7-3 A leak detection sensor connected to a service line at the meter , 85

7-4 A leak detection node contained in a hydrant cap, 85

7-5 A typical communications network, 85

7-6 Typical node layout on a transmission main (top) and a node installation in an above-ground pedestal (bottom), 87

7-7 Tethered leak detection tool, 89
7-8 Free-swimming leak detection device, 90
7-9 Gas tracer leak detection method, 91
7-10 Basic diagram of low-voltage conductivity circuitry, 94
7-11 Steps in data capture, analysis, and display with satellite-based leak detection, 95

8-1 Images from a CCTV inspection of a 21-in. steel water main, with close-up of an exposed joint (right), 98
8-2 Example of a side scan camera deliverable, 99
8-3 Digital video and hydrophone sensor head (left); field deployment of tethered sensor head into live transmission main (right), 100
8-4 Laser scan with better than 2-mm accuracy exported to AutoCAD for manipulation and analysis for areas of concern (i.e., 1, 2, and 3), 103
8-5 Examples of crawling ROVs with CCTV cameras, 103
8-6 Image from a high-definition camera with zoom capability on crawling ROV, 104
8-7 Image from a high-definition camera on crawling ROV in dry conditions, 104
8-8 Tethered CCTV insertion setup (top left), tethered CCTV setup through a fire hydrant (top right), and operation of a long-range system with parachute to propel the sensor with flow (bottom), 106
8-9 Real-time, forward-looking sonar data showing a grate in a flooded pipe, 107
8-10 Typical cross-sectional view of profiling sonar data, 107
8-11 Flooded tunnel visualization using multiple multibeam sonars, 108
8-12 Advantages of combining profiling and forward-looking sonar sensors, 109
8-13 Sample output from an underwater laser scan of a horseshoe-shaped conduit, 110
8-14 A laser scan of a dewatered conduit using a stationary laser platform, 110

9-1 Demonstration of physical entry in various pipe diameters, 117
9-2 Electronic data collection during physical entry inspection, 122

10-1 AV testing history and progression timeline, 126
10-2 Validation database, 127
10-3 Typical setup for an acoustic velocity measurement, 128
10-4 In-pipe acoustic velocity system, 130

11-1 Example of a free-swimming tool, 137
11-2 Example of a high-resolution, large-diameter tool, 137
11-3 Example of a self-propelled crawler, 138
11-4 Example of a handheld tool, 138
11-5 Example of a manually winched tool, 139
11-6 Example of a hydrant-launched, water-propelled tool, 139
11-7 Simple probe for remote field testing, 140

12-1 MFL field in unlined pipe, 151
12-2 MFL field in lined pipe, 152
12-3 B-H curve, 152
12-4 Typical MFL ILI tool in CML pipeline, 154

12-5 Test data in ½-in. pipe with ½-in. mortar lining, 155
12-6 Isolated pipeline defects, 156
12-7 MFL data from severely corroded pipeline and removed pipe showing corroded area, 156
12-8 MFL view of circumferential corrosion and exposed pipe showing circumferential corrosion, 157
12-9 Color contour C-scan of blind test defects, 158
12-10 Graphical depiction of pipeline defect anomaly definitions, 159

13-1 Pre-stressed concrete cylinder pipe, 164
13-2 Break rates for PCCP manufactured in different eras, 167
13-3 Typical tools and equipment used in visual and sounding inspections, 168
13-4 Example of longitudinal crack surrounded by hollow, 169
13-5 Electrical currents induced by time-varying magnetic field, 171
13-6 Technician pushing an electromagnetic inspection cart through a PCCP, 172
13-7 Tethered and untethered robotics platforms equipped with electromagnetic inspection technology, 172
13-8 Example of pipe excavation for external electromagnetic inspection, 174
13-9 Illustration of sonic/ultrasonic testing of PCCP, 176
13-10 Fiber optic cable anchored to pipe in dry deployment (left) and deployed into in-service pipeline in wet deployment (right), 178

14-1 Catastrophic break due to bad weld at intersection between 30-in. and 36-in. pipes, 187
14-2 Pressure testing for LADWP, 189

15-1 Electromagnetic scanning of pipe within keyhole excavation, 198
15-2 Keyhole excavation with pavement repair, 195
15-3 Acoustic velocity testing on a Los Angeles trunk line, 196
15-4 Inserting an RFT tool into the hydrant lateral of a shallow-buried main, 197
15-5 Photo showing insertion of the RFT tool during assess-and-fix demonstration in Phoenix, 199
15-6 Fracture analysis of a PVC pipe break, 199

16-1 Condition assessment data displayed in Google Earth, 204
16-2 Summary table, 205

A-1 Pipe wall analysis using magnetometry, 208

This page intentionally blank.

Tables

1-1 Typical condition assessment project items, 5
1-2 Simple cost-benefit analysis for condition assessment of a 10-mile, large-diameter pipeline over 10 years, 6
1-3 Simple cost-benefit analysis for condition assessment of a cast iron pipeline over 10 years, 6

3-1 Available inspection technologies for pressure pipe, 24

4-1 2018 AWWA Benchmarking Survey Distribution System Integrity results (combined leaks and breaks/100 mi/y), 36
4-2 Example criteria and weightings for LoF scoring, 38
4-3 Example criteria and weightings for CoF scoring, 39
4-4 Example criteria for GIS-based CoF scoring, 40

6-1 Common in situ field and laboratory tests for spot assessment, 60

7-1 Parameters for sounding equipment, 81
7-2 Parameters for listeners, 81
7-3 Comparison of leak detection correlators, 82
7-4 Parameters for correlators, 82
7-5 Parameters for leak loggers, 83
7-6 Parameters for remote correlating leak detection systems: distribution systems, 86
7-7 Parameters for remote leak detection systems: transmission mains, 87
7-8 Parameters for tethered devices, 89
7-9 Parameters for free-swimming devices, 91
7-10 Parameters for gas tracer method, 92
7-11 Parameters for ground-penetrating radar, 93
7-12 Parameters for thermography, 93
7-13 Parameters for low-voltage conductivity testing, 95
7-14 Parameters for satellite-based leak detection, 96

8-1 Inspection platform/method options, 102
8-2 Operating parameters for tethered CCTV inspection, 105
8-3 Relative costs of internal inspection methods and technologies, 112

9-1 Complementary testing techniques performed during physical entry inspection, 123

11-1 Available tools using electromagnetic technology, 141

12-1 Definitions of pipeline anomalies, 158
12-2 Estimated performance specifications, 160

14-1 Dispersion scenario and appropriate testing method, 188
14-2 Test pressure limits, 190

Preface

Water main condition assessment currently ranks among the most important subjects to water utilities, and interest in this subject is rapidly growing. As the average age of their infrastructures increases, utilities are increasingly challenged to maintain levels of service while keeping water affordable to all. Condition assessment helps utilities meet this challenge by identifying more precisely where money is best spent, leaving in place pipelines that have adequate integrity and preventing the unnecessary failures of others.

The assessment of buried pipelines is never simple, and water mains are particularly challenging. They are difficult to access, and concerns exist regarding water contamination and the need to maintain water service for sanitation, commerce, and fire protection. Because water main condition assessment is relatively new, there's a scarcity of authoritative, peer-reviewed material to guide utilities. Instead, utilities often are left to rely on the claims of inspection companies, some of which are not well established. This manual fills an important need, providing a comprehensive overview of the many methods that can be used to assess water mains.

Methods range from desktop studies to leak detection to full-length, high-resolution scans from inside the pipes. The choice of method and the interpretation of results are both art and science, with decisions based on economics and risk tolerance. This manual provides the technical information utility managers and engineers need to make informed decisions, along with practical information about how methods can be deployed.

For assessment of most pressure pipelines, a multistep approach is recommended. By starting with simpler, less-expensive examinations, then progressing if necessary to more expensive techniques, resources can be leveraged. Assessment is meant to answer questions about questionable pipes, where a decision is needed that affects service or cost. Assessment is usually less useful when applied to pipes that are known to be in very good or very poor condition. Pipes may be distribution or transmission mains, where the consequences of a break range from mundane to intolerable. In either case, with condition assessment, decisions can be made more confidently and the likelihood of a poor decision reduced. Risks are better managed.

This manual is largely organized around the various inspection methods, with chapters also devoted to program management. The physical inspection techniques focus on detecting degraded materials—corroded metal in particular. Readers should keep in mind that other causes of main failure, including ground movement (pipe bending), pressure surges, casting defects, poorly made welds, and other joint problems may be difficult (if not impossible) to detect in advance of failure.

It is important to understand the limitations of the various inspection methods: what sizes and types of defects might be missed; whether the method has been verified by independent, third-party verifications (dig-ups); and where blind spots exist. Several methods, for instance, have difficulty finding defects near pipe joints. Irregularities such as riveted seams or heavy scales may produce "noisy" data. No method is perfect, and no inspection method finds all possible defects. The methods and techniques included in this manual have been reviewed by the Water Main Condition Assessment Committee and found to be worthy of consideration, but not all have been proven through years of use and *independent* testing. Inclusion in this manual does not imply AWWA or committee endorsement of any particular method or the companies that provide it. Utilities are encouraged to start with the information presented in this manual, consult other publications, seek referrals, and ask hard questions before investing.

The applicability of these methods varies depending on pipe material. Some methods apply to all types of pipe, while others apply more narrowly to metal or reinforced concrete pipes. The focus is also on water mains (large and small). It is usually not economical to assess small service laterals with these methods. Chapter 3 provides guidance regarding which methods can be used for which types of pipe and also regarding selecting and procuring services. This manual focuses on methods that are commercially available in North America, and because the field is rapidly evolving, other methods are expected to be introduced in the next few years. A few of these are listed in Appendix A.

Unlike the other chapters in this manual, Chapter 13 addresses one specific material: prestressed concrete cylinder pipe (PCCP). PCCP has been particularly problematic due to its propensity to fail without warning and in a catastrophic manner. As a result, much attention has been paid to the assessment of PCCP, and many tools and techniques specific to this type of pipe have been developed. Some of these tools are now being applied to other types of material, particularly bar-wrapped steel cylinder concrete pipe.

Through the publication of this manual, AWWA hopes to provide practical information that water utilities will use to better manage their systems. AWWA also seeks your feedback regarding this manual and other ways to advance this important discipline. This marks the start of a continuous, dedicated effort to collect and disseminate reliable information on water main condition assessment. Your participation in this process is welcomed and encouraged.

—Water Main Condition Assessment Committee
Dan Ellison, PE – Chair

Acknowledgments

This manual is a product of the Water Main Condition Assessment Committee. Special thanks to the following individuals who devoted considerable time to the endeavor.

Committee Chair and Editor: Dan Ellison, HDR, Ventura, Calif.

Chapter Lead Authors:

Chapter 1—Jeff Leighton, Portland Water Bureau (retired), Portland, Ore.

Chapter 2—Kurt Vause, Anchorage Water and Wastewater Authority, Anchorage, Alaska

Chapter 3—Dan Ellison, HDR, Ventura, Calif.

Chapter 4—Celine Hyer, Arcadis, Tampa, Fla.

Chapter 5—Jeff Giddings, HDR, Omaha, Neb.

Chapter 6—Mersedeh Akhoondan, HDR, San Diego, Calif.

Chapter 7—Ahmad Habibian, CDM Smith, Fairfax, Va.

Chapter 8—Andi Corrao, infrastructureMD, San Diego, Calif.

Chapter 9—Derek Wurst, Black & Veatch, Walnut Creek, Calif.

Chapter 10—Frank Blaha, Water Research Foundation, Denver, Colo.

Chapter 11—Ricardo Hernandez, Metropolitan Water District of Southern California, Los Angeles.

Chapter 12—Chris Macey, AECOM, Winnipeg, MB, Canada, and Rod Jackson, CH2M Hill (now Jacobs), Sacramento, Calif.

Chapter 13—Bethany McDonald, Black & Veatch, Phoenix, Ariz.

Chapter 14—Andrew Romer, AECOM, Orange, Calif.

Chapter 15—Dan Ellison, HDR, Ventura, Calif.

Chapter 16—Nathan Faber, San Diego County Water Authority, Escondido, Calif.

Chapter Coauthors:

David Conner, AECOM
Andi Corrao, infrastructureMD
Tracy DeLa Torre-Evans, Seattle Public Utilities
Kris Embry, Hibbard Inshore
Nathan Faber, San Diego County Water Authority
Chris Garett, PICA
Duane Gilles, Evansville Water & Sewer Utility
Mark Grabowski, Electro Scan
Michael Grahek, Los Angeles Department of Water & Power

Sylvia Hall, PE, Sylvia Hall Engineering
Margaret Hannaford, San Francisco Public Utilities Commission
Yakir Hasit, American Water
Phil Hoyt, Electromechanical Technologies
David Hughes, American Water (Retired)
Rod Jackson, CH2M Hill (now Jacobs)
Dave Johnston, Echologics, Division of Mueller
Peter Kraft, The Confluence Group
David Kroon, Aegion (Corrpro)
George Kunkel, Kunkel Water Efficiency Consulting
Kevin Laven, Echologics, Division of Mueller
Jeff Leighton, Portland Water Bureau (retired)
Joanna Line, City of Calgary
John Marciszewski, Echologics, a division of Mueller
Ashan McNealy, Pure Technologies
Mike McReynolds, Brown and Caldwell
Paul Meschino, Utility Service Group
Kenneth Morgan, KCM Consulting Services
Trent Nedens, Ballard Marine Construction
Rasko P. Ojdrovic, Simpson, Gumpertz, and Heger
Elly Perets, Utilis
Noy Phannavong, V&A Consulting Engineers, Inc.
Emma Quail, Pure Technologies
Annie Raven, InfraPlan
Martin Roubal, Rock Solid Group
Piero Salvo, GAME
David Spencer, HDR
Allison Stroebele, Pure Technologies

Other Contributors/Technical Reviewers

Scott Arnold, Aegion (Fyfe)
Graham Bell, HDR
Tom Bowman, JD7
Roy Brander, City of Calgary
Keith Browning, Orlando Utilities Commission
Jim Eggen, City of Joliet
Matt Gaughan, Plus Six Engineering
Chuck Hansen, Electro Scan
Gordon Henrich, Pipeline Integrity Technology Associates
Jeremiah Hess, Portland Water Bureau
Dave Kozman, Hammerhead Trenchless
Daniel Krywyj, Aquam Pipe Diagnostics
George Kunkel, Kunkel Water Efficiency Consulting
Bryon Livingston, Black & Veatch

George Mallakis, T.T. Technologies
John C. Matthews, Trenchless Technology Center
Mo Najafi, University of Texas
Chris Nastally, Great Lakes Water Authority
Fred Pfeifer, Washington Suburban Sanitary Commission
Dave Russell, Russell NDE Systems
Russell Titus, New Jersey American Water
Jon Turner, Phoenix Civil Engineering
Travis Wagner, Pure Technologies
Cameron White, Pure Technologies

AWWA Staff: Sue Weikel, Jim Siriano, Michelle Sheehy, Mindy Burke, Janet Greifinger, Jennifer Santini, Willadee Hitchcock

Dedication: This manual is dedicated to Phil Hoyt, who pioneered the use of magnetic flux leakage for water pipeline inspections. Phil contributed greatly to this manual and, sadly, passed away before seeing its completion.

This page intentionally blank.

Chapter 1

The Benefits of Condition Assessment for Water Mains

Jeff Leighton, Chapter Lead, Portland Water Bureau (retired)
Nathan Faber, San Diego County Water Authority
Yakir Hasit, American Water
David Conner, AECOM

This chapter describes generally what condition assessment is and what it can accomplish. In brief, condition assessment involves the deployment of methods to ascertain the physical condition of a water main. By performing condition assessment of water mains, a utility can

- estimate the likelihood that a water main may continue to provide satisfactory service, both now and in the future;
  - help determine the remaining service life of the water main;
- make better decisions regarding main renewals;
  - allow some water mains to remain in service longer;
  - prevent some pipeline failures from occurring by intervening sooner;
  - make decisions more confidently (with less chance of error);
- find active leaks; and
- tailor renewal projects and methods to better match the conditions of the mains.

DEFINITION OF CONDITION ASSESSMENT

The emphasis of this manual is on methods of pipeline condition assessment to identify physical condition. Condition assessment may be defined as the identification of the likelihood that an asset will continue to perform its required function. As part of condition assessment, data and information are collected through direct and/or indirect methods and then analyzed to determine the physical characteristics of the pipe and how they may impact the pipeline's likelihood that it will leak, break, or otherwise fail to perform. Example characteristics include current or future structural, water quality, and hydraulic status of an individual pipe, segment, or collection of similar pipes, among other characteristics.

Condition assessment may be performed in the field, via desktop, or both. The important objective is to do it, update it, and improve upon it as needed. Field condition assessment involves direct and indirect observations of the asset and its environment to determine and document its condition. Desktop condition assessment relies more heavily on existing data and institutional knowledge to make the same determination using design documents, staff knowledge, information systems, industry experience, and other resources to determine or approximate the condition of the pipelines without viewing them physically. Beyond these efforts, condition assessments may use more advanced study and testing to more fully identify pipeline condition.

Some examples of how a condition assessment is used include

- to identify loss of integrity, and that water is leaking—water loss may be observed or detected indirectly through acoustic methods;
- to identify loss of structural competence or weakening of the pipe or that the wall thickness is diminished—wall loss may be established through a variety of methods;
- to find evidence of liner or coating failure—may be visually observed; and
- to recognize other conditions of concern, e.g., pipe is unacceptably out of round.

Condition Assessment and Monitoring as Part of a Risk-based Asset-Management Strategy for Pipes

A fundamental activity for any water utility is to determine the risks associated with asset failure. Understanding the risk of asset failure and determining an acceptable level of risk for the utility allows the balancing of conflicting goals of minimizing lifecycle costs of assets versus delivering the stipulated levels of service (LoS). Risk analysis is used to understand the cause, effect (consequence), and likelihood of events adverse to attainment of LoS; managing such risks to an acceptable level; and providing an audit trail for the management of risks.

Mathematically, risk from a failure can be expressed as the product of the consequence of the failure (CoF) and the likelihood of the failure (LoF):

$$\text{Risk} = \text{CoF} \times \text{LoF}$$

Risk analysis is used to rank assets by their risk of failure and to identify high-risk assets (i.e., assets with a risk of failure above an acceptable level of risk). In assessing risk, CoF and LoF are quantified separately, and the results can be multiplied to calculate the risk-of-failure score of a specific asset.

An asset is considered to be failing if it cannot, or does not, provide the requisite LoS. For water mains, this failure is measured by physical condition, hydraulic performance, and quality of water. Thus, when determining the LoF of a pipeline, these three factors should be assessed, with the physical condition being the most prominent one.

The risk associated with each water main can be quantified by multiplying the CoF and LoF values to arrive at a relative risk score. Often a scale such as 1 to 5 is applied to both CoF and LoF. Applying the risk equation to each of the water mains under consideration allows for a comparison of the mains by their relative risks. Because the scoring is somewhat subjective, the absolute risk score value itself is not as important as is the relative risk ranking of a main. Listing the mains by their relative risk score usually makes clear which mains are posing the highest risk, which are posing the next highest risk, and so on. The relative risk score is then used to help identify options for risk mitigation.

When computing LoF scores, water utilities typically have limited condition information on all the water mains in their system. This information might consist of break and leakage data for some of the mains, field condition analysis data on a few critical mains, etc. Furthermore, some asset data used might be dated, incomplete, or unreconciled, and therefore less reliable. For the remainder of the mains, most of the information that is used is typically based on institutional knowledge and the application of cohort analysis techniques. Thus, when the risk score is calculated for all the water mains in a system, these scores need to be vetted, especially for the high-risk mains for which mitigation measures such as rehabilitation or replacement might be considered. Before making any significant investments for mitigation, it is important to verify the validity of the data used by conducting further condition assessment activities on these mains.

Condition assessment is a fundamental element of the strategic asset-management effort. It helps capture and make available timely information including pipeline condition and pipeline attributes (e.g., identifiers, sizes, materials, and uses) that are used to implement an asset management program. Condition assessments can be recurring and may be improved upon in future iterations. Initial condition assessment will provide key elements of asset management to establish minimum initial awareness of asset condition and needs, as well as a baseline of these for future comparisons. Condition assessments are also useful to

- show and realize progress in and benefits of asset management;
- engage staff members and attract their participation, contributions, and buy-in; and
- increase value in the short-term and long-term pipe-renewal investments for the asset-management effort.

The accomplishments of condition assessment contribute to the organizational aspects of service levels, strategy, risk, prioritization, process efficiency and effectiveness, quality information systems and tools, performance monitoring, and others.

THE BENEFITS AND COSTS OF CONDITION ASSESSMENT

Benefits of Condition Assessment

The overarching benefit of condition assessment is the ability to gather data for risk-based decision making. Condition assessment data can therefore lead to better and more defensible decisions and the reduction of risks associated with water mains. Most agencies strive to achieve efficient operations through main break avoidance and efficient renewal planning.

Condition assessment data can lead to break avoidance by identifying and proactively repairing pipes that are likely to fail. Avoiding breaks can also lead to customer benefits including more reliable service, reduction of unplanned outages, and positive customer perceptions. The financial benefits of break avoidance include lower repair costs for planned repairs, less damage to surrounding infrastructure, and more defendable claims from private parties stemming from pipe failures (Martel 2014). Social impacts of pipe breaks include traffic and business disruptions. Also, there are environmental benefits to avoiding breaks, such as habitat damage, unnatural sediment buildup, and negative water quality impacts from discharged water.

Another key benefit of condition assessment data is efficient renewal planning. Condition assessment provides data to support and justify better planned pipe-restoration projects. Utilities can focus renewal efforts on pipelines that need repair rather than basing decisions on general characteristics that may not correlate to actual condition such as age. In many cases, condition assessment efforts reveal that most of the pipeline is in good condition. This information helps utilities avoid unnecessary pipe replacements. If only a small part of the pipeline needs repair, utilities can focus their repair efforts on these areas and benefit from extended use of the sections of the pipeline that are in good condition. Understanding the repair needs for the long term (10–20 years) helps the utility plan and execute those projects and schedule projects in a sustainable way with regard to resources. Condition assessment data can also help define project boundaries and bundle similar projects to reduce community impacts and lower project mobilization costs. The condition assessment data also support the utility's decision for a renewal project to both internal and external stakeholders. In many cases, the utility can physically show the data to stakeholders to effect understanding and context of the repair need. Finally, condition assessment data can confirm or even extend the assumed life of a pipeline. For example, a condition assessment effort might indicate that an old pipeline previously thought to need replacement is in good condition and can continue to provide service for decades longer.

Costs of Condition Assessment

The costs of condition assessment vary greatly depending on the type of condition assessment, location, size, requirements, access, environment, assessment methods, and system complexities. This section is meant to serve as a guide to the various costs that may apply. The utility owner should determine whether a given cost applies and develop an estimate. Table 1-1 provides a list of important items that the utility should consider including in a condition assessment project.

Cost-benefit examples. To illustrate the justification for condition assessment, two simple cost-benefit comparisons were developed: one for a large-diameter case and one for a small-diameter case. We are using two techniques, magnetic flux leakage (MFL) and remote field technology (RFT), for illustrative purposes only. The case for using other techniques can also be made. The point is that the cost of condition assessment (for these two techniques and for others) can be justified. The first analysis looked at a future MFL condition assessment project. The comparison is for a 10-mile, steel, large-diameter (84-in.) pipeline over a 10-year term. The costs were based on past project costs and included preparing, draining, accessing, and filling the pipelines. The assumption was made that as a result of this condition assessment, four locations of significant corrosion would be identified and repaired. For prevention of a pipeline leak, the low repair costs were based on fixing four leaks over 10 years, each at a cost of $300k, which includes a pipeline shutdown, draining, access, repair, filling, and property damage/claims due to leaks. The high estimate of $6 million is due to an estimated $1.2M for each leak in additional property damage and claims impacting several homes, structures, businesses, and other utilities. Proactive repairs based on an MFL condition assessment would have a benefit of avoided repair and damage costs ranging from $1.2M to $6M over a 10-year period. The life extension benefits are based on the estimated cost of constructing a replacement pipeline, which would last 100 years, with new construction costs of $88M in a rural area and $120M in an urban area.

Table 1-1 Typical condition assessment project items

Vendor/Contractor Costs	
1	Mobilization
2	Work execution
3	Analysis
4	Reporting
Owner Costs	
1	Planning/coordination (meetings, schedules, correspondence, plans, calculations, permits, forms, and notices)
1-1	Customers (service interruptions, temporary service, water storage)
1-2	Public (notification, accommodations)
1-3	Regulatory agencies (water discharge, disinfection, work permits, traffic permits)
1-4	Contractors (scheduling and coordination)
1-5	Other needed pipeline maintenance, repairs, and/or upgrades
2	Preparation
2-1	Safety (pipeline isolation, communication, emergency protocols, lockout/tagout, ventilation, trench shoring, equipment)
2-2	Access to facilities/pipeline (vehicles and equipment) and access inside pipeline (valve removal, excavations, flange removal)
2-3	Water discharge/erosion control
2-4	Equipment availability/functionality (e.g., generators, pumps, air blowers, trench shoring, trench plates, discharge filtering, portable restrooms, cranes, small tools)
2-5	Valve testing/functionality/configuration
2-6	Contracts/agreements for water service and/or water supply
2-7	Material needs (gaskets, nuts, bolts, paint, grease, coatings, couplings, paving, sand, gravel)
3	Implementation
3-1	Pipeline access, draining, excavation, ventilation
3-2	Visual inspections
3-3	Planned and unplanned pipeline maintenance, repairs, and/or upgrades
3-4	Contractor/vendor condition assessment work execution
4	Protocols
4-1	Communication and progress reports
4-2	Safety (equipment, rescue teams, lockout)
4-3	Traffic control
4-4	Pipeline disinfection and testing
4-5	Equipment and materials (see above)
5	Data integration
5-1	Review
5-2	Field verification
5-3	Analysis
5-4	Integration with existing data

To calculate the avoided costs based on a life extension for 10 years, these replacement costs were depreciated (straight-line) over a term of 10 years. Table 1-2 illustrates the simple cost-benefit comparison.

The second analysis looked at a 12-in. (30.48-cm) diameter cast iron pipeline condition assessment project using RFT. The comparison is for a 4,000-ft (1,119-m) section of pipeline over a 10-year term. The simple cost-benefit comparison is provided in Table 1-3.

Table 1-2 Simple cost-benefit analysis for condition assessment of a 10-mile, large-diameter pipeline over 10 years

	Description	Assumptions	Low Cost*	High Cost†
Costs	MFL condition assessment	Total project	$1.5M	$2.0M
	Steel repairs	Internal steel patches (4 total)	$30k	$30k
Benefits	Prevention of a pipeline leak	Damage and repair costs (4 leaks total)	$1.2M	$6.0M
	Avoided depreciation cost	10-year life extension	$8.8M	$12.0M
		Net cost savings:	$8.5M	$16.0M
		Benefit/cost ratio:	6.7	9.0

MFL—magnetic flux leakage
* Rural area
† Urban area

Table 1-3 Simple cost-benefit analysis for condition assessment of a cast iron pipeline over 10 years

	Description	Assumptions	Low Cost	High Cost
Costs	RFT condition assessment	Total project	$100k (access available)	$300k (excavation and pipe modification for access)
	Repairs	3 repairs	$120k	$150k
Benefits	Prevention of a pipeline leak	Damage and repair costs (3 leaks total)	$120 (low consequence)	$840k (high consequence)
	Avoided depreciation cost	10-year life extension	$400k	$400k
		Net cost savings:	$300k	$790k
		Benefit/cost ratio:	2.4	2.8

RFT—remote field technology

CONDITION ASSESSMENT AS PART OF AN ASSESS-AND-FIX STRATEGY

The objective of assessing the pipe is to understand the condition of the pipe. Some interest exists in the industry to use condition assessment as part of a structured water main rehabilitation program. The concept is that using condition assessment tools to measure wall loss can aid in the maintenance of small-diameter, less consequential aging iron pipes, especially those that have never been lined. Water Research Foundation Project 4473 described and demonstrated this approach: using nondestructive examinations (remote field technology was the applied technology in the demonstrations) to help the utility select a lining system based on the assessed pipe condition (Bell 2015). Lining systems vary from nonstructural methods applicable to mains with little corrosion to semistructural to fully structural liners to carry the pipe loads where significant corrosion and wall loss has occurred.

This strategy is commonly applied to nonpressurized sewer pipes, where rehabilitation methods are selected based on video assessment results. Several factors make this more challenging for pressured water pipes: the high cost of assessments, the difficulty of putting assessment tools into the pipe, taking mains out of service and associated customer impacts, the concern about water quality when tools are introduced into the pipe, the difficulty of interpreting the assessment data, and the lack of utility experience with the approach or the availability of guidance. Nevertheless, it is both feasible and often cost-effective. This will be described in more detail in Chapter 14 of this manual.

ADDITIONAL CONSIDERATIONS WHEN USING CONDITION ASSESSMENT

Concerns and Challenges of Condition Assessment

Condition assessment challenges include providing reliable data that have a sufficient level of detail and locational accuracy to make informed decisions. In many cases, the utility's expectations do not align with what the contractor hired to conduct a condition assessment is planning on providing. A number of factors, including calibration of the equipment, the accuracy of the data being provided, and the assumptions based on pipe characteristics, can impact the reliability of the data.

Funding and/or technical issues may limit a utility's condition assessment options. Operational complications due to shutting pipelines down, achieving proper isolation, and recharging and disinfecting the line can be a challenge. In some cases, the utility may not have the required system information such as pipe plans, manufacturing details, locations, and/or repair information. At the end of a project, the utility may not understand the significance of the findings and know how to act on the data. In many cases, the data may not be compatible with existing data formats, and it may be difficult to integrate them. On some projects, the utility may not be fully aware of the limitations of the data, and these are not disclosed before the project.

The contractor sometimes assumes important details about a utility's system, or the utility does not properly convey system information such as pipe interconnections, isolation valves, previous repairs, and/or operational parameters. For example, there should be a clear understanding of when and where the contractor would be allowed to operate water system valves. Additionally, the data interpretation may be difficult for the contractor to convey and difficult for the utility to understand.

Predictive Failure Analysis

The knowledge and data derived from condition assessment may be used in predictive analytics to forecast probable time ranges of likely pipeline failure. Also, trending of the existing pipelines and trends of historical similar pipelines may be used in this effort. A variety of modeling and statistics may be leveraged to develop predictions of pipeline failure. By applying the knowledge of current condition captured in the assessment, these endeavors help to construct a typically more realistic curve of pipeline failure rather than assuming a linear degradation over time based only on such attributes as type, material, age, type of use, etc. They also help to customize the curves more to the particular organization and its pipelines because they leverage actual information rather than relying solely on industry generalizations.

These analytics may be advanced to enable scenario analyses based on operation, maintenance, performance, and other drivers to forecast pipeline failure risk, timing, and modes, often in almost real-time dashboard environments. Geospatial and hydraulic modeling software may be helpful to access, manage, and communicate the information of the predictive or real-time analyses. Modeling can also be used to quantify the CoF, helping the utility to prioritize which pipes should have priority for condition assessment.

Predictive analytics is especially useful in situations where current condition is unknown, inaccessible, and/or cost prohibitive to capture and is especially useful for pipe networks. Various studies are exploring predictive analytics, and standards such as

International Standards Organization (ISO) 24516, "Guidelines for Management of Assets of Water Supply and Wastewater Systems," are giving recognition to the field.

The costs associated with predictive analytics include the capture, management, and manipulation of various pipeline data and asset condition information and could also include a myriad of other information depending on the intended scope of the predictive effort. Costs also include the staff time involved in the predictive modeling and could increase with the use of geospatial and hydraulic modeling software, though such software may already be on-hand.

REFERENCES

Bell, G.E.C. 2015. The Assess-and-Fix Approach: Using Non-Destructive Evaluations to Help Select Pipe Renewal Methods. Denver: Water Research Foundation.

Martel, K. 2014. Best Practices for Water Utility Legal Protection and Claims Management from Infrastructure Failure Events. Denver: Water Research Foundation.

Chapter 2

Building Support for a Condition Assessment Program

Kurt Vause, Chapter Lead, Anchorage Water and Wastewater Authority
Jeff Leighton, Portland Water Bureau (retired)
Kenneth Morgan, KCM Consulting Services
John Marciszewski, Echologics, a division of Mueller
David Hughes, American Water (retired)

An effective condition assessment program requires funding, which requires support from the public, managers, and other stakeholders. This chapter discusses ways to build support by

- making a business case for condition assessment;
- developing effective methods for accounting for condition assessment costs;
- developing an appropriate budget; and
- gaining support from policy makers, the public, and other stakeholders through effective communications.

MAKING A BUSINESS CASE FOR CONDITION ASSESSMENT

Business case evaluations are key components of advanced asset-management programs at many water utilities. Condition assessment business cases, in particular, are important for water systems where the value of the infrastructure is under-appreciated. Like any capital investments, business cases can be developed to justify the expenditure of funds to perform condition assessment. Given the wide array of degradation rates for buried water pipelines, best practices demand that business cases be prepared (Committee Report 2016).

The purpose of condition assessment may be to validate, estimate, justify, or improve the quality of decisions regarding water assets (Vanrenterghem-Raven and Vause 2015). Whereas desktop or statistical means may be used to assess condition and performance levels of specific assets, physical condition assessment can augment such measures. For example, when using desktop analysis of existing utility data to statistically determine the degradation of pipe cohorts, insufficient data may be a limiting factor; in such cases, condition assessment is used to validate other methods used to describe asset performance. If knowledge regarding a specific pipe segment is desired, or if utility-specific data are lacking and statistical methods are therefore impractical, condition assessment can be employed to evaluate the performance of asset groups that behave in a similar manner.

Condition assessment is also used to validate other methods used to describe asset performance. Where proper application of desktop analysis suggests it is cost-effective to invest in rehabilitation or replacement of investments, field condition assessment can be used to justify the right amount of expenditures and to determine specific assets to renew. Finally, condition assessment results can improve upon desktop or statistical analysis of asset performance, wherein the cost of condition assessment is justified by the degree of risk mitigation achieved, and lead to improved targeting of capital (e.g., identifying pipe needing repairs, renewal, or replacement). For example, a major break on a large pipe may not justify replacement of miles of pipe if condition assessment reveals that only a few sections merit attention.

Business Cases for Condition Assessment

Business cases for condition assessment are typically predicated upon demonstration that the expense associated with the effort offsets otherwise planned investments, reduces risk costs, or validates asset integrity to develop reliable estimates of asset deterioration over time. Several utilities have documented returns on investment of 400–800 percent when implementing an "assess-and-address" approach, in which condition assessment is used to find and target specific pipe defects rather than replacing an entire main. Additionally, several utilities have been able to use capital, as opposed to operational, budgets (Coghill et al. 2014) for pipeline condition assessment and repair since the projects are extending the useful life of assets that may have previously had zero book value based on their age.

One example that illustrates the value of using a business-case approach to condition assessment includes the experience of Anchorage Water Wastewater Utility (Vanrenterghem-Raven and Vause 2014). The utility used statistical methods to conduct in-depth failure analysis of its pipe inventory by accessing power analytical tools. With its own internal capabilities, the utility developed correlations between pipe types and break rates and created its own plan for replacement of sections of its distribution system. One section of the system contained high-consequence, small-diameter metallic pipe situated in a railroad yard terminal (Figure 2-1).

One section of the distribution system surrounding the rail yard was recommended for replacement using the utility's economic replacement model, which used a triple-bottom-line failure costing technique in conjunction with statistical failure modeling. Several failures of that section were the primary drivers for replacement. Nearby sections of the distribution system were also recommended for replacement following engineering review, despite limited break history, based on proximity and similarity (same vintage, material, and installation) to the pipe section originally recommended for replacement. To justify these expenditures, additional detailed condition assessment was recommended to confirm the replacement need.

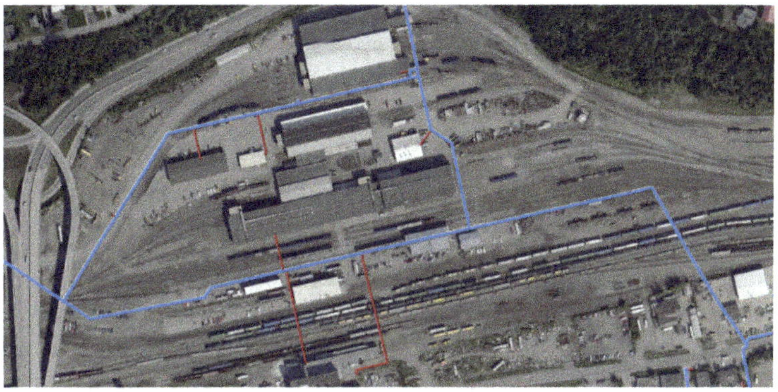

Source: Vanrenterghem and Vause 2014
Note: Blue lines are water mains, and red lines are service laterals.

Figure 2-1 Aerial view of Anchorage, Alaska, railroad yard

Given the high consequence of pipeline failure (CoF) within the rail yard, which included direct costs of interrupting the Alaska Railroad's commercial activities (involving freight shipments and passenger tourism), plus the cost of emergency repairs, product loss, potential fines, and flooding impacts, a high-resolution condition assessment technology was selected. Deployment required close coordination between the utility, its contractor, the Alaska Railroad, and water customers in the vicinity of the rail yard. Scheduling of the work was complicated by the need for the railroad to maintain train operations in the project area. The costs were considerable, because inspected pipe included unlined cast-iron sections with moderate tuberculation, and excavations were needed for insertion of the inspection equipment. Burial depths exceeded 10 ft, consistent with Anchorage's practice of placing mains very deep for freeze protection.

Results showed remaining pipe wall averaged between 86 and 96 percent of original pipe wall thickness in inspected sections, with few localized defects requiring repairs. Pipe integrity over the full length of inspected sections revealed no leaking sections. As a result of this high-resolution inspection, the utility made the decision to forego much of the planned replacement and to focus on lower-cost repairs to limited sections of the pipe. The savings were substantial.

The costs of assessment included contract inspection and analysis, laboratory pipe coupon testing, and pipeline access construction, plus internal costs for engineering, project management, and distribution system operations support. Expenses totaled more than $500,000.

Benefits included a reduction of capital project scope as compared to the original engineering analysis and replacement recommendations, which totaled $2.11 million. It also included avoiding a future annual depreciation expense of newly installed pipe, estimated at nearly $39,000. In terms of value of all benefits, total annual savings to the utility were calculated to be $166,000 per year, creating a payback period of 3.4 years for the condition assessment expense.

Moreover, the utility calculated the net present value of 20 years of risk reduction achieved by performing conditon assessment and an altered project scope. Risk reduction was achieved by avoiding potential expenses of interrupting an important economic activity (commercial activities of the Alaska Railroad). The triple-bottom-line risk cost was estimated at $3.7 million. This expense reflected the costs of emergency repairs, social impact (railroad impact), and associated environmental impacts from mainline breaks. Net risk reduction savings were determined to be $1.5M after counting for condition assessment and the revised project scope, demonstrating value gained by this approach. Comfortable with these results, the utility now uses similar practices for its program of pipeline replacement and has received support from its governing board in subsquent requests for program funding.

ACCOUNTING TREATMENT: WAYS TO ACCOUNT FOR CONDITION ASSESSMENT COSTS

Utilities interested in using condition assessment are often faced with challenges in getting staff agreement over proper means to account for the expenses to be incurred. If condition assessment work can be capitalized, it is often easier to fund because capital improvement funding is considered an investment rather than a mere cost. The argument for capitalization is that condition assessment can create value—adding new life expectancy to mains that have been fully depreciated. Unfortunately, this argument is often a hard sell.

The treatment of expenses associated with condition assessment requires adherence to generally accepted accounting principles. As a rule of thumb, for municipally owned utilities, all preliminary costs leading to the activation of a pipeline condition assessment project, including desktop analysis of breaks and leaks, landslide and soil movements, ground deformation, over-the-line survey on impressed current potential measurements, soil resistivity, other pipe corrosion factors, and other studies identifying sections of pipe with high consequence or high likelihood of failure, should be expensed, not capitalized. Routine pipeline inspection is typically expensed.

Certain private water companies may have access to state programs that allow for rate increases for non–revenue producing investments to replace aging infrastructure. Therein, latitude may exist for condition assessment to be capitalized. As one example, the Distribution System Investment Charge (DSIC) program was started in Pennsylvania in 1996. Benefits include more efficient and timely investment of capital, significant progress in replacing aging infrastructure, enhanced service quality, reduction of water leaks, rate shock avoidance, and others. Asset rehabilitation or replacement are capitalized items under the DSIC program.

Capitalization of condition assessment is often acceptable if the following conditions are met:

- the likelihood of replacement or improvement of the assessed section of pipeline becomes *reasonably certain;*
- the future improvement or replacement qualifies for capitalization; and
- the estimated total pipeline condition assessment project cost exceeds a utility-designated threshold—say, $100,000.

Reasonably certain is defined as follows.

- Preliminary findings support that a qualified capital improvement will need to occur. Preliminary findings support includes a desktop analysis identifying sections of pipe that have a high consequence or high likelihood of failure based on leak repair history, site-specific inspection records, ground deformation, soil resistivity, and other pipe corrosion/degradation factors, as well as condition assessment results.
- The capital project for replacement/improvement is included in the capital improvement plan, and funding is available.
- The activation of the conduit condition assessment project has been approved by management for creation, and it is expected that more than 50 percent of the assessed pipeline will be replaced or improved through some level of rehabilitation.

Pipeline condition assessments will be capitalized as an ancillary cost of the future pipeline improvement or replacement.

Qualifications for capitalization of the replacement or improvement will typically include meeting both of the following criteria:

- total improvement or replacement meets the utility's capitalization threshold for infrastructure, which is typically set at an amount of not less than ($5,000–$10,000), and
- full replacement of a section (typically one full length) of pipeline or an improvement that extends the accounting useful life of a section of pipeline by at least 50 years.

Examples of such improvement include

- cement-mortar lining (extends pipe life approximately 30–50 years) and
- structural lining (extends pipe life approximately 50–75 years).

While individual utilities may have different thresholds and different rules, at least one application of these principles dictated that the pipeline condition assessment cost would be expensed if it was later determined that the pipeline condition did not result in replacement or improvement of more than 50 percent of the accessed pipe.

BUDGETING FOR A SYSTEMATIC CONDITION ASSESSMENT PROGRAM

Finding sufficient budget to execute a systematic condition assessment program without proper planning and foresight is often difficult. Often, condition inspection expenses are unknown during budgeting processes as the exact infrastructure to be inspected is undesignated or inspection plans are not yet fully developed. As a result, it is necessary to anticipate, plan, and formulate a systematic program that includes appropriations for operations, in addition to specific projects involving condition assessment.

To this end, leading utilities are developing methods to determine what amount and form of assessment is needed to provide the confidence required to meet their level of service (LOS) targets and execute their capital improvement program and/or repair budgets. Condition assessment planning and budgeting is thus becoming an integral component of master planning processes at water utilities.

Through development of master plans, risk assessments (assessing which assets warrant evaluation) and cost-benefit assessments of condition-inspection technologies are performed. With knowledge of asset types and sizes to be inspected, appropriate technologies are identified and typical costs of an inspection program developed. Combined with risk assessment documenting the consequences of failure, properly quantified in terms of risk-cost exposure, an assessment of cost-effective technologies is produced. Cost-effectiveness is judged by comparing the amount of expense for condition assessment against the benefit obtained (reduction in risk cost). Through this mechanism, budgets can be developed and programmed for future action.

Such budget actions can create a cycle by which systemic condition assessment becomes part of a utility program of asset management improvement. For example, with noninvasive acoustic technologies, utilities may realize a real return on condition assessment costs by simultaneously determining the structural integrity of pipe segments and accurately locating leaks, all without breaking ground or disrupting service, at an assessment cost that provides economic benefit in excess of the financial (direct and indirect) consequences of asset failure.

One practice in use today is the application of the inverted condition assessment pyramid (Figure 2-2) to make decisions about specific buried assets or cohorts (Laven 2014). With the goals of mitigating risk and maximizing capital investments, the process starts with a desktop study and only continues far enough down the pyramid until sufficient

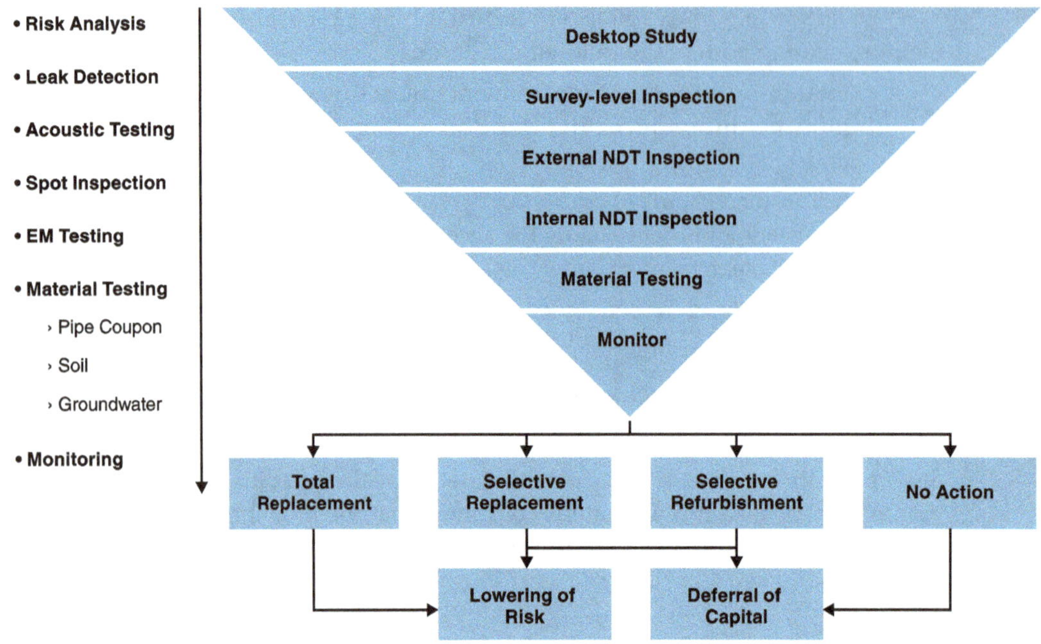

Adapted with permission from Laven (2014)
EM: electromagnetic, NDT: non-destructive technique

Figure 2-2 Progressive condition assessment and resulting actions

information is obtained for a given asset. The pyramid approach helps utilities make an informed choice for repairs, replacement, rehabilitation, or leaving the asset uninspected for a few more decades.

It may not be necessary to follow each of the steps in the pyramid after the initial desktop assessment of risks. For pipes with low risks, no further steps may be needed in the short term. For some high-risk pipes, a utility will find it prudent to move to a detailed condition assessment method and skip over the interim steps. (The selection and application of appropriate condition assessment methods are further discussed in Chapter 3.)

With such an approach, a utility must consider whether condition assessment on specific buried assets provides sufficient information that enables a cost-effective program to be developed over an entire asset inventory. Therefore, the linkage between condition assessment and master planning continues to be the subject of additional research, with the intent being to formulate a generalized roadmap for utility use.

GAINING SUPPORT FROM POLICY MAKERS, THE PUBLIC, AND OTHER STAKEHOLDERS

With a solid business case, gaining support from policy makers, the public, and other stakeholders is possible. Such support is necessary to secure funding and foster public understanding and support for a sustainable program of condition assessment. Often, condition assessment is outside the realm of typical utility-funded activities, so outreach and education are key tactics to employ.

Tactics also may follow other utility investment proposals, such as focusing efforts on value, risk management, and finances. The Utility Branding Network for Water and Waste Water Agencies (2016) identifies the need to create a "compelling argument for investment," which includes

- a "program title" that succinctly describes the intended objective;
- rules, levels of quality, or desired achievements that are considered acceptable;

- a problem statement that best expresses inability to meet a current level of service, standard, or compliance with a new standard or regulation;
- a method or description of how the problem may be solved—this may be in the form of a specific investment, process, or program;
- communication of capital and operating expenses of the proposed program and whether current budgets and rates already allow the necessary funding;
- an answer to the question "why now?" and why the proposed time frame for action is appropriate—this should include the logic behind the timing;
- sharing considered alternatives in advance to demonstrate an objective, thorough, and transparent process; and
- a critical evaluation of the consequences of failing to implement the proposed program, usually expressed in terms of increased risk or cost. The consequences of not proceeding must be understood and made part of the discussion.

With respect to arguments for a structured, sustainable condition assessment program, improved planning incorporating the results of condition assessment accomplishes two key facets of utility operations. First, long-term replacement plans have been demonstrated (Vause et al. 2013) to be optimized and investment expenditures minimized, as compared to capital plans that do not employ condition assessment. Second, use of condition assessment to improve planning efforts can better align utility performance to desired service standards and help identify which asset types must receive attention in appropriate time frames. These two key results often help policy makers and stakeholders appreciate and understand the importance of supporting assessment programs.

One example of a structured, sustainable condition assessment program is that of the City of Calgary, Alberta. As described by Ellison et al. (2018), for more than 15 years the city has used data from in-pipe, electromagnetic scans to help select water mains for rehabilitation or replacement. Each year, the city scans a small portion of its system, wherein the sections scanned are selected using several criteria that are important to the city's unique characteristics. This technique has proven most useful for water mains where only a few breaks have occurred and the mains are considered candidates for anode retrofits. If evaluations reveal pipes in a badly deteriorated condition, planned retrofits can be abandoned and mains scheduled for replacement. In addition to these assessments, Calgary also applies data from more than 100,000 soil resistivity readings to infer condition of pipe below ground. This structured program has had the effect of reducing the number of main breaks by one-half over the 15-year period, producing annual savings in repair costs that more than cover the cost of the inspection and retrofit program. At the same time that breaks have been reduced, the main replacement program has also been substantially reduced.

In building a business case for the condition assessment effort, particularly for large-diameter pipe, a utility can expect the following:

- a reduction in the negative impacts of the pipe failure on customers and the environment if not identified proactively;
- a planned approach to needed repair instead of a reactive one;
- a reduction in the costs to complete the repairs and to restore service;
- less impacts on customers because of planning as opposed to a response to an emergency;
- a reduction in the amount of water loss as a result of a planned repair versus a main break; and
- an increased awareness of specific weak areas that alert the utility to be prepared for repair, renewal, or replacement work.

In addition to the direct costs of a planned repair (based on a condition assessment) versus an emergency response to a break, there are other associated costs and/or expenses, such as

- claims against the utility for property damage and/or loss of revenues from businesses;
- a reduction in the utility's anticipated revenues as a result of the break repair; and
- the expense of requiring additional staff time, equipment, and materials to isolate a greater area for the leak repair as opposed to the lesser amount when the job is planned properly.

For a government-owned utility, costs incurred by the public (i.e., the owners of the utility) are to be considered as well. These include costs associated with traffic delays and loss of service, even if no claims are ever filed. Total costs may be much higher than one would imagine. In the case of large-diameter pipe, a study of 30 main failures found a geometric mean of $500,000 total cost per failure, with the largest costs associated with flooding damage (Gaewski et al. 2007). The time required to close valves appeared to greatly influence costs.

Both the direct and indirect costs of asset failures have been described in greater detail by the Water Research Foundation (Raucher et al. 2017).

SUMMARY

To build support for a condition assessment program, keys to success include

- establishing a solid business case for condition assessment that demonstrates the expense associated with the effort offsets otherwise planned investments, reduces risk costs, or validates asset integrity;
- understanding accounting treatment of expenses and developing a plan in advance of the work that conforms to applicable standards of the utility; and
- anticipating, planning, and formulating a systematic program that includes appropriations for operations, in addition to specific capital projects involving condition assessment.

Taking these steps will improve planning efforts to better align utility operations to desired service standards and help identify which asset types must receive attention in appropriate time frames. A program of this type can also optimize investment expenditures as compared to capital plans that do not employ condition assessment.

REFERENCES

Coghill, M.R.; Eaton, G.A.; and Faber, N.D. 2014. A Long-Term Commitment to Pipeline Infrastructure: Implementing, Funding and Delivering the San Diego County Water Authority's Asset Management Program. Proc. Pipelines 2014 Conf., ASCE, Portland, Ore.

Committee Report: Business Practices to Help Utilities Better Manage Assets. 2016. *Journal AWWA*, 108(1):61–67.

Ellison, D.; Spencer, D.; and Brander, R. 2018. Leveraging Data from Non-Destructive Examinations to Help Select Ferrous Water Mains for Renewal. Denver: Water Research Foundation.

Gaewski, P.E.; Tata and Howard, Inc.; Blaha, F.J.; and AwwaRF. 2007. Analysis of Total Cost of Large Diameter Pipe Failures. Denver: AWWA Research Foundation.

Laven, K. 2014. Acoustic Structural Pipeline Testing. Proc. 2014 Netherlands Asset Management Symposium, Almere, Netherlands.

Raucher, R. 2017. Managing Infrastructure Risk: The Consequence of Failure for Buried Assets. Denver: Water Research Foundation.

Utility Branding Network for Water and Waste Water Agencies. 2016. Making a Compelling Argument for Investment. http://utilitybranding.net/pdfs/UBN-Compelling-Argument-Flyer.pdf (accessed Dec. 13, 2018).

Vanrenterghem-Raven, A. and Vause, K. 2015. Estimation of Service Lives: Using a Combination of Statistical Analyses and Field Work. Presentation at AWWA 2015 Annual Conference and Exposition, Anaheim, Calif.

Vanrenterghem-Raven, A. and Vause, K. 2014. Optimizing Physical Condition Assessment Dollars Using Statistical Results. Presentation at 2014 INFRA Congress, Center for Expertise and Research on Infrastructures in Urban Areas (Canada), Montreal, Quebec.

Vause, K.; Raucher, R.; Reynolds, H.; Vanrenterghem Raven, A.; Miller, C.; and Helgeson, T. 2013. Use of Infrastructure Replacement and Renewal Planning Tools. Workshop Presentation at 2013 AWWA Annual Conference and Exposition, Denver.

Chapter 3

Planning a Condition Assessment Project or Program

Dan Ellison, Chapter Lead, HDR
Tracy DeLa Torre-Evans, Seattle Public Utilities
Noy Phannavong, V&A Consulting Engineers, Inc.
Kris Embry, Hibbard Inshore
Kenneth Morgan, KCM Consulting Services
Duane Gilles, Evansville Water & Sewer Utility

This chapter provides an overview of several important subjects to consider when planning a condition assessment project or program, including

- what level of condition assessment is needed,
- how to choose an assessment method for a specific pipeline or location,
- what to consider when planning field work,
- what to consider in communicating the project to the community, and
- how to contract for condition assessment.

LEVELS OF CONDITION ASSESSMENT

The assessment of water mains should be approached like a physician approaches a physical exam. She does not immediately order a full-body scan but rather starts with simple, inexpensive tests (blood pressure, reflexes, etc.), then moves to more expensive examinations if warranted. Also, multiple tests are performed. No single test provides all the data.

The more serious the possible consequences, the more intensive the examination. Examination of a young, seemingly healthy individual may consist of a few basic tests every several years, whereas as an older person with chronic problems may require many tests performed more frequently.

Similarly, the condition assessment of a pipeline is usually best done progressively, with the results of one inspection determining what additional inspections are appropriate. Higher-risk pipelines are given more attention, and as their condition deteriorates, the assessments become more frequent.

In determining the level of assessment to perform, there are generally four variables to be considered.

1. **The degree of inspection:** Are you ascertaining the general condition of the pipe, or do you want to find small defects?
2. **The inspection coverage:** How much of the pipe is to be assessed?
3. **Pipeline access:** How easy is it to get near or in the pipe with various tools? Can the pipe be shut down and dewatered? How long can it be out of service?
4. **The cost of the assessment:** A utility's willingness to invest in condition assessment should be in some proportion to the perceived risk and the overall value of the asset. An expensive assessment is harder to justify for assets where the replacement cost is relatively inexpensive.

As shown in Figure 3-1, the higher the degree of inspection and the greater the amount of pipe to be inspected, the higher the cost of the assessment.

Additionally, the cost of pipeline inspection is greatly affected by economies of scale, the competitiveness of the marketplace, how easily the pipeline can be accessed, and the development costs associated with the technologies. Unit costs will be lower when longer

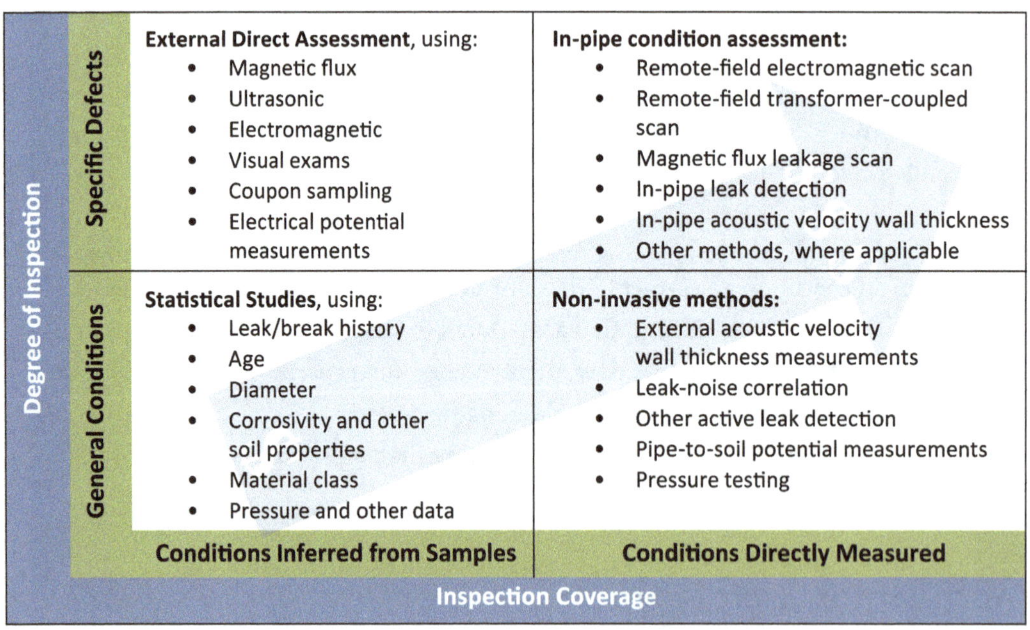

Source: HDR. Modified from Laven et al. (2011)

Figure 3-1 The tradeoffs between costs, coverage, and degree of inspection

run lengths or several pipelines can be inspected with a single mobilization of an inspection technology company.

CHOOSING AN ASSESSMENT METHOD

Pipeline Accessibility Is a Major Consideration

As much as we would like a method that finds all the small defects with no disruption of operations and with only a minor expenditure, technologies have limitations. In general, a tool capable of a high-resolution examination will need to be relatively close to the pipe wall surface. This means excavating to expose the pipe or inserting a tool (or person) inside the pipe.

The ability to access a pipeline is therefore a major consideration. If tools (or personnel) will be used inside the pipe, suitable openings will be required, and operations may be affected. In some cases, existing hydrants, blowoffs, and manways can be used, but in other cases a tool-launching facility will need to be constructed. Some in-pipe tools require a dewatered pipeline, whereas others can be propelled through the pipe using normal water flow. High-resolution assessments typically require tools that are tight-fitting to the pipe bore, so a relatively large access opening is required. If the pipe can be dewatered, some of these tools can be inserted as disassembled pieces through existing manways and assembled from inside the pipe.

For some in-pipe tools, particularly the high-resolution devices, the pipe will need to be cleaned with pipeline pigs before tool insertion. If the pipe is choked with scale (tuberculation), the pigging effort can increase by a magnitude. The pigging is to ensure unimpeded and relatively smooth travel of the tool and also helps to verify that obstructions that block tool passage probably do not exist.

In many cases, multiple contractors may be involved with the inspection. In addition to the pipe inspection company, contractors are often used for constructing access ports and to assist with the pipe cleaning.

Obstructions Must Also Be Considered

The passage of a large tool can be easily blocked by butterfly valves, plug valves, and tight-radius bends. Even when gate valves are present, reducers on both sides of the valves may block passage of the tool. A thorough examination of record drawings and interviews with operations staff is required to determine the potential for such obstructions.

Even small tools can be impeded or lost if the inspection plan does not adequately account for how and where water will flow during the inspection. The flow velocity must be adequate to propel the tool up positive slopes and keep the tool from straying off course. The presence and operability of lateral valves is therefore important.

Progressive Examination Often Saves Money

Often, money can be saved by starting with simpler, less expensive examinations before implementing more expensive tests. Information gathered from the simpler examinations may be sufficient to determine whether the pipe is in relatively good or poor condition.

The general progression for water main condition assessment is as follows, but steps are often omitted, combined, or rearranged as appropriate. Seldom is every step employed.

Step 1. Risk assessment to determine which water mains warrant the most attention

Step 2. Desktop analysis of drawings, soils reports, break/repair history, operational conditions, and staff interviews to determine the general likelihood of a break

Step 3. Site reconnaissance and inspection planning (accessibility, traffic, other utilities)

Step 4. Indirect assessments, including:

- Corrosivity survey
- Corrosion-activity measurements (close-interval survey)

Step 5. Leak detection surveys (external or internal)

Step 6. Low-resolution direct assessments

- Acoustic-velocity measurements of pipe stiffness
- Detection of stress anomalies through analysis of electromagnetic fields

Step 7. Excavations/external exams where leaks, anomalies, or corrosion hot spots are identified or suspected, including:

- High-resolution scanning tools (ultrasonic, magnetic flux leakage, remote-field/broadband electromagnetic)
- Extraction of soil, mortar, and groundwater samples for laboratory examinations
- Electrical measurements to detect corrosion, coating defects, and bi-metallic couples
- Grit-blasting and pit-depth measurements
- Visual and photographic exams

Step 8. Internal physical entry or robotic visual (closed-circuit television, CCTV) inspections, including various scanning and inspection techniques

Step 9. Internal, full-length, high-resolution scans (e.g., magnetic flux leakage or remote-field/broadband electromagnetic)

The results of earlier steps are used to determine the necessity for subsequent actions and to target specific areas. Also, because no single method finds all defects, using multiple methods helps ensure a more complete understanding of pipe condition.

For some pipes, pressure testing may be a suitable alternative to Steps 3 through 7. As discussed in Chapter 14, pressure testing can find rather quickly and economically large weaknesses in large-diameter pipes (areas that may fail catastrophically) but may not find all the small weaknesses.

Opportunity Assessments

Because water mains are not easily accessed, when opportunities to observe either the interior or exterior of a water main occur, they should be considered. Among these opportunities are the following.

- **Repairs.** When mains are repaired, data should be gathered regarding the mode of failure, the properties of the materials, and the condition of the pipe. Photos should be taken, and GPS coordinates should be recorded. Examples of suggested repair reports can be found in various AWWA and Water Research Foundation publications.
- **Service taps.** When pipes are tapped for new service connections, not only is the pipe observed, but coupon samples are often extracted, which can be measured and tested.
- **Failure analyses.** Unexpected, premature, and unexplained failures often merit an engineering/scientific (i.e., "forensic") assessment, involving laboratory testing of the failed materials and a review of the design and installation of the pipeline and

its operations. Because the results can lead to litigation, these investigations may need to be thorough, detailed, and unbiased.

- **Service interruptions.** A depressurized, dewatered pipeline allows for interior observations that are otherwise more difficult and often not as effective when the pipe is filled. Opportunities arise from tie-ins and other construction work, as well as repairs.

Opportunity assessments can sometimes be a fraction of the cost of other assessments because they take advantage of the traffic control, excavation, shoring, permits, and other logistics that are already in place. However, as with any spot assessments, the results must be interpreted with caution, as the conditions revealed may not be indicative of conditions at other locations along the pipeline.

AVAILABLE TECHNOLOGIES

The materials with which and the way a pipe is manufactured determines which methods are applicable (see Table 3-1). For metallic pipes, a larger range of technologies exists, making use of the electrical, magnetic, and electromagnetic properties of these materials. On the other hand, fewer methods are available for plastic materials (polyvinyl chloride [PVC] and high-density polyethylene [HDPE]), where failures generally stem from small defects that are difficult to find in advance.

The last column in Table 3-1, "Typical Recommended Approach," shows a suggested sequence of steps that may be used for each material.

Leak Detection Methods

Leak detection methods apply to essentially all pipes, but their degree of success often depends on the material and size of the pipe. Large, massive pipes such as concrete attenuate (diminish) sound more than small metallic pipe, so acoustic leak detection is more difficult. Likewise, sound attenuates on plastic pipe more than on stiffer materials.

Leak noise correlation is a widely practiced method that can be used to help pinpoint leaks using two points of contact. It has been widely used on small-diameter, metallic pipes, but systems are now available that accurately locate leaks on nonmetallic and large diameter mains.

In-pipe leak detection tools can overcome some of these limitations, such as for large-diameter pipes with few appurtenances, and can find very small leaks. These in-pipe tools can also detect gas/air pockets.

Several other leak detection tools and methods can also be used, as discussed in Chapter 7.

Visual and Other Physical Inspection Methods

Physical observations are also useful for many types of pipes and conditions. As described in Chapter 8, visual observations can be accomplished using CCTV cameras mounted on robotic crawlers, push-in probes, miniature submarines, and other platforms. These inspections often include laser and sonar instruments for measuring various physical conditions. For larger pipelines, physical entry ("manned" inspections) by trained personnel can be useful—not just for observation, but for performance of various physical tests from inside the pipe, as described in Chapter 9.

A word of caution: looks can be deceiving. Although CCTV is widely used in wastewater and other gravity pipes, the defects that cause pressure pipes to fail are often not easily observed. In water mains, CCTV can determine the presence (or absence) of lining

Table 3-1 Available inspection technologies for pressure pipe

Method	Soil Corrosivity Surveys	Acoustic	Remote Field Electromagnetic	Ultrasonic	Magnetic Flux Leakage	Broadband Electromagnetic	Other	Typical Recommended Approach
Description	Various electro-magnetic, electrical, and laboratory methods characterize the corrosivity of soils, detect/measure corrosion activity, and assess effectiveness of corrosion protection/cathodic protection	Acoustic velocity: pipe wall stiffness is calculated from the speed of sound transmission Acoustic monitoring: alerts and pinpoints the location of wire breaks	Changes in electromagnetic signals indicate broken wires, corrosion pits, and changes in wall thickness and stress anomalies	Reflection of sound waves is used to measure the thickness of various types of materials. Tool must have direct contact or liquid coupling with material being measured	Changes in magnetic fields are used to detect corrosion pits and other defects. Tool must be at constant, close distance from pipe wall	Changes in electromagnetic signals indicate corrosion pits and changes in thickness. Scanner held near pipe, but works through coatings, linings, and scale	Sampling of pipes for various physical tests Manned entry for visual and sounding (delamination testing) Petrographic (microscopic) examinations of concretes and mortars	GENERAL APPROACH (all pipe types): 1. Statistical analysis of available data 2. Risk prioritization (likelihood and consequence of failure) 3. Records review (leak/break repairs, drawings, specs, reports, soil info) 4. Site reconnaissance (accessibility, traffic conditions, other utilities) 5. Inspection planning (shutdowns, bypass, permits, alternatives) 6. Leak detection and/or field condition assessment inspection

(continued)

Table 3-1 Available inspection technologies for pressure pipe (continued)

Method	Soil Corrosivity Surveys	Acoustic	Remote Field Electromagnetic	Ultrasonic	Magnetic Flux Leakage	Broadband Electromagnetic	Other	Typical Recommended Approach
Asbestos Cement (AWWA C402)	Assess potential for AC (concrete) deterioration (pH and sulfates)	Acoustic velocity can detect gross deterioration	n/a	n/a	n/a	n/a	Testing of samples: • Phenolphthalein stain • SEM/EDS • petrography	1. Tests of opportunity samples from repairs and service taps 2. GIS mapping of soil data, breaks, and condition assessment data 3. Targeted condition assessment of high-consequence pipes
Prestressed Concrete Cylinder Pressure Pipe (AWWA C301 and C304)	Assess potential for metal and concrete deterioration Detect active corrosion	Acoustic monitoring for detection of wire breaks. Acoustic velocity can be used for prioritization of other assessments.	Used to detect broken wires	n/a	n/a	External and internal spot assessments	Internal sounding to detect delamination Internal visual (manned entry or CCTV) External direct assessment Petrographic analysis of mortar/concrete	1. Risk analysis based on pipe type, manufacturer, wire type, year of manufacturer, corrosivity 2. Manned entry/sounding (if feasible) 3. Electromagnetic scanning 4. Acoustic monitoring
Non-Prestressed Concrete Pressure Pipe (AWWA C300, C302, and C303)	Assess potential for metal and concrete deterioration Detect active corrosion	n/a	Emerging method to detect broken bars and cylinder corrosion	n/a	n/a	External spot assessments. Emerging method for internal scanning	Internal sounding to detect delamination Internal visual (manned entry or CCTV) External direct assessment Petrographic analysis of mortar/concrete	1. Corrosivity survey 2. External direct assessment where corrosion risk is highest 3. Manned entry examination 4. Electromagnetic scanning

(continued)

Table 3-1 Available inspection technologies for pressure pipe (continued)

Method	Soil Corrosivity Surveys	Acoustic	Remote Field Electromagnetic	Ultrasonic	Magnetic Flux Leakage	Broadband Electromagnetic	Other	Typical Recommended Approach
Ductile Iron Cast Iron (AWWA C150 & C153)	Assess corrosivity to iron Monitor corrosion activity	Acoustic velocity can detect gross deterioration	Detailed internal scan of pipes and external spot assessment. Works with cement mortar and tuberculation	Used for external spot assessments	Internal scanning of non-CML lined pipes. External spot assessments.	External spot assessments. Emerging method for internal scanning	Petrographic analysis of mortar	1. Corrosivity survey 2. Remote field electromagnetic
Steel (AWWA C200)	Assess potential for metal and concrete deterioration Monitor corrosion activity	Acoustic velocity can detect gross deterioration	Used for detailed internal scan of pipes. Works with cement mortar and tuberculation.	Used for external spot assessments	Internal and external scanning of both CML and non-CML pipes	External spot assessments. Emerging for internal scanning.	Petrographic analysis of mortar	1. Corrosivity survey 2. Pipe-to-soil potential; cathodic protection assessment 3. Remote field electromagnetic or magnetic flux leakage
Copper	Assess potential for metal deterioration	n/a	Used for detailed internal scan of pipes	n/a	n/a	Potential for detailed scan of pipes	Forensic examinations of failed pipes Electrochemical noise monitoring	1. Evaluate construction methods and standards 2. Evaluate soil corrosivity 3. Forensic exams of failures
Plastic Pipes (HDPE – AWWA C906) (PVC – AWWA C900)	n/a	n/a	n/a	n/a	n/a	n/a	Forensic examinations of failed pipes, using laboratory and mechanical tests	1. Review of drawings, specs, and inspection records 2. Forensic examination if early or frequent failures have occurred

Source: HDR

AC: asbestos concrete, CCTV: closed-circuit television, CML: cement mortar lines, GIS: geographic information system, HDPE: high-density polyethylene, n/a: not applicable, PVC: polyvinyl chloride, SEM/EDS: scanning electron microscopy with energy-dispersive spectroscopy

and the degree of scaling, but the integrity of the pipe itself will generally not be discernible. Moreover, even when the pipe is visible, the degradation is often hidden. Graphite and other corrosion products obscure the deterioration of iron pipes. The deterioration of asbestos cement (AC) and concrete pipes is similarly hidden.

Inspection Limitations

It is important to understand not only the capabilities but also the limitations of the various inspection methods: what sizes and types of defects might be missed; has the method been verified by independent, third-party verifications (dig-ups); where are the blind spots? Several methods, for instance, have difficulties finding defects near pipe joints, and irregularities such as riveted joints may produce noisy data. In many cases, the amount of damage that is detected will be understated, because no method detects every type of defect.

PLANNING OF FIELD WORK/COMMUNICATIONS

Depending on what methods are employed, planning can range from simple to complex. In many cases, work occurs in roadways, so safety is a concern, and traffic control and associated permits may be required. Generally, the larger the pipe, the greater the effort required to plan and execute the work.

If tools will be inserted within potable water mains, a plan is necessary for how water quality will be protected. Only tools designated for potable water should be used, and they will need to be disinfected before entering the system. It is likely that sediment in the pipe will be stirred up, which will cloud the water. If a tuberculated main is being scanned, greater thought is required; disturbance of these scales can lead to longer-range "dirty water" issues and possible concerns about release of contaminants that may be concentrated in the scales. Most of these issues are solved by isolating the main before the work is performed and flushing the main after. This may include closing service lines so customers do not receive turbid water and possibly flushing the services upon completion of work. In general, the processes and procedures should be similar to how the utility handles main repairs.

A bypass piping system can be installed as a way of providing continued water supply to customers while the assessment is underway, preventing worry about water contamination. Such systems are typically laid along both sides of the road with provisions for vehicle passage at driveways and other crossings. The systems are disinfected and fully tested before they are placed into service. AWWA M28, *Rehabilitation of Water Mains,* provides additional information.

Customer notifications are essential for any work where shutdowns will occur or where water turbidity is a possibility. When properly informed of the purpose of the work, most customers will appreciate that the utility is performing the assessment and will often find the "high-tech" work interesting. Communications may be by letters, community meetings, and (of course) e-mails and the Internet. Providing customers with phone numbers, e-mail addresses, and other means of communicating their concerns and questions is also important. Providing the customer with free bottled water may also be helpful in generating goodwill.

For large pipes, pipes in busy streets, and non-tethered tool insertions, the inspection company, the engineer, and various utility managers should meet to discuss the various risks and determine appropriate contingency plans. While the work is not dramatically different from work that the utility performs routinely, roles and responsibilities need to be understood. As the adage goes, "failing to plan is planning to fail."

Logistical Considerations for Performing Internal Inspections

When employing any technique where a tool is inserted inside the pipe, several logistical issues must be considered.

- **Shutdown, isolation, and dewatering of pipeline.** One of the most important considerations for selecting and performing internal inspections is whether the main can be shut down, isolated, and dewatered. This will dictate what methods can be practically deployed to inspect the pipeline interior. Associated concerns include water discharge requirements and permits.
- **Access points into pipeline.** Another significant consideration is whether there are access ports, flanges, manways, or launching stations on the pipeline. Having to install new access points adds significant cost to the condition assessment. It is also important to know where the access points are located and how accessible they are. Associated concerns include traffic control, shoring, and permit acquisition.
- **Pipeline operating conditions.** For inspection of operational pipes, operating pressure and flow velocity are important considerations. Pressures above 200 psi can be problematic for some tools. For devices that are propelled by water flow, velocities that are too low can cause the tool to stall, and velocities that are too high can result in poor data collection.
- **Constraints due to pipeline configuration.** The pipeline's physical configuration (size, alignment, potential obstructions) has to be considered in relation to the tool or technique that is proposed.
  - *Pipe size.* Generally, physical (manned) entry is practical down to about 36-in. diameter for pipelines. Entry into smaller pipes is possible, but there are increased safety concerns. Manned-entry inspections must always be carefully planned and thoroughly reviewed.
  - *Pipe bends and pipe slope.* Changes in direction both horizontally and vertically can hinder internal inspections by limiting the distance covered from an access point. Cable drag increases significantly at pipe bends. Normally, a tethered tool can be safely pulled through no more than about 300° of bends (e.g., three 90° elbows). Sharp, short-radius bends can block tool passage or damage cables.
  - *Pipe length.* Long travel distances may make manned entry difficult, and tethered tools will always be limited by cable length as well as the number of bends and operating conditions (flow and pressure).
  - *Obstructions.* Internal obstructions created by valves, appurtenances, debris, and scale buildup (tubercles) can limit the length of internal inspections or manned entry. For example, butterfly valves and other non–full-bore valves may be impassible for large tools and pose a risk that equipment will get stuck. Cleaning can address the issue of scale buildup, but cleaning requires an additional step and perhaps an additional shutdown. Complaints of discolored water may occur if the cleaning is not well managed.

Health and Safety Considerations

As with all utility activities, health and safety issues are the most important considerations. If the inspection cannot be performed safely, without risks to consumer health, then it needs to be reconsidered.

Many utilities may be able to rely on their existing standard processes and procedures, but exactly who is responsible for what must be determined in advance. Often staff from several departments need to be brought together with one or more outside contractors to finalize the plan and ensure regulatory compliance. Among the issues to be considered are

- the closing of line and lateral valves to isolate the main and customer impacts;
- lock-out/tag-out procedures;
- disinfection of tools and materials that contact water or are inserted in pipes;
- traffic control, barricades, shoring, and other worker protection;
- confined-space entry procedures—air monitoring, personnel retrieval, and permits; and
- disinfection, flushing, and testing of mains and laterals post inspection.

Generally, inspection tools used in wastewater mains should not be used in water mains. Most inspection companies understand this and segregate tools accordingly, but the question should always be asked.

CONTRACT GUIDELINES

Unlike video inspections performed on sewer mains, water main condition assessment is never a commodity. Seldom are similar services offered by competing companies. Even where two companies offer similar-looking work (e.g., electromagnetic scanning), it is important to understand the differences in what each company provides, both in terms of the final product and how much field support will be required. Tools that appear to be similar can in fact be very different in their ability to detect pipe defects. Equally important is the amount of skill and effort that is devoted to interpretation of results. For most technologies, a very well-trained technician is needed to take the data produced by the device and translate them to useful information.

In soliciting work, a "request for proposals" (RFP) is thus a more appropriate means of obtaining services than a "request for bids." If selection is done on a "low-bid" basis, the utility is likely to get the least value as well as the lowest cost. An RFP should include

- a description of what the utility is trying to accomplish;
- project requirements, including
  - the contractor's responsibilities and
  - the utility's responsibilities;
- constraints, including permits;
- proposal requirements; and
- evaluation criteria and method of selection.

Examples of RFPs can often be obtained from consultants, other utilities, and the technology companies themselves.

In evaluating proposals, the greatest attention needs to be paid to the experience and expertise of the company, its technology, and the information it will provide. Some technologies have been introduced to the market without being thoroughly tested, whereas other technologies have undergone years of independent tests, including tests by the Water Research Foundation and the US Environmental Protection Agency (USEPA), where pipes have been exhumed to verify both positive findings (i.e., the presence of defects) and negative findings (i.e., the absence of defects). Proving that a technology can find a defect is less important than proving that it won't miss a defect.

While most utilities have a "can-do" attitude and very smart technical people, it may be helpful to also employ a consultant who is well versed in this work. Websites and company literature often do not provide enough information to distinguish the apple from the orange and know which one is better for a particular situation.

SUMMARY

Many issues must be considered in planning a condition assessment project, including selecting the methods that are most appropriate for the pipelines and implementing them effectively. Method selection hinges on the goals of the assessment, how readily the pipe is accessed, what obstructions might exist within the pipe, and how much the utility is willing to pay. Investments in condition assessment should relate to the perceived risk of a pipeline and the expected benefit of the assessment.

As with all field work, safety is the prime consideration, and for work on potable water systems, maintaining sanitary conditions is a foremost concern. Effective customer communication is a priority; good communication builds and maintains support for the program and helps protect the utility's reputation.

Because water main condition assessment is both relatively new and rapidly evolving, utilities often lack the necessary knowledge to confidently plan and execute these projects. While this manual helps to fill the gap, there is no substitute for experience, which is available through the AWWA Water Main Condition Assessment Committee.

REFERENCE

Laven, K.; Kong, X.; and Nguyen, V. 2011. Recent Advances in Pipe Wall Assessment Technology. ASCE Pipelines Conf., Seattle.

Chapter 4

Desktop Condition Assessment

Celine Hyer, Chapter Lead, Arcadis
David Spencer, HDR
Annie Raven, InfraPlan
Peter Kraft, The Confluence Group

A desktop assessment generally precedes the field inspections that are described in later chapters. The term "desktop" means a review and analysis of available records and documents, including system data, repair records, drawings, geotechnical reports, failure analyses, and other *existing* data. Because these data come from utility files, a desktop assessment is relatively inexpensive compared to the collection of new data. The collection of new data can also be disruptive, involving service interruptions and other community impacts. Desktop condition assessments help determine what additional data and assessments are worthwhile.

The purposes of desktop assessments are to

- calculate and forecast system performance (break rates and service levels);
- estimate the remaining useful lives of assets and determine appropriate renewal rates;
- determine which mains represent the highest risks to the utility based on likelihood and consequence of failure; and
- determine which of the high-risk mains are candidates for physical condition assessment.

OVERVIEW

This chapter discusses important concepts for performing a desktop condition assessment, including the following:

- the data needed to perform a valid and meaningful desktop assessment;
- how to define a pipeline asset and cohorts through statistical analysis to support replacement planning;
- what levels of service are and how they support replacement planning;
- performing a risk assessment using pipe-specific characteristics to develop a "likelihood of failure" score and a "consequence of failure"; and
- developing performance decay curves and using them to create short- and long-term replacement plans and estimate remaining useful lives.

A desktop condition assessment often entails using existing pipeline characteristics, operational data, and failure data to better understand the likelihood of failure (LoF) of a pipe or a group of pipes with similar behavior compared to others (i.e., a cohort). It is a relative assessment that is frequently used by itself to identify long-term financial renewal needs, or to select pipes for short-term replacement programs. Desktop assessments are particularly helpful for distribution system mains, where there is sufficient historical break data history and where the cost to physically assess individual mains is high relative to their replacement costs. Desktop assessment for transmission mains, where break data are generally very limited, becomes an extrapolation from other pipe characteristic variables or from other systems.

For distribution mains, often there is a sufficient break history to perform meaningful utility-specific statistical analysis. Variables such as material, age, diameter, and geographic location typically provide strong enough statistical correlations to forecast future breaks. For large transmission mains, where few if any break records may exist, it becomes more difficult to assess condition from the desktop. At best, one can gauge the general risk based on industry-wide experience, taking into consideration the material, its age, stress levels, and environment. Where risk and cost/benefit analyses support it, the desktop assessment is then supplemented with direct condition assessments as described in other chapters.

DATA CLEANUP AND ASSESSMENT

A variety of data can be used as part of the desktop assessment to determine both the likelihood of failure and the consequence of failure. The data need to be available for all pipe segments that are analyzed. In addition, the data may need to be cleaned before the analysis to get the best results.

Data Sources

Any data that describe the physical condition of a pipe and the factors that contribute to its degradation are of interest for desktop condition assessments. These include

- physical characteristics—material, pressure class, wall thickness, cathodic protection, coating, lining, year of installation, year of abandonment, diameter, length, etc.;
- operating conditions—pressure, surge events, water quality, customer complaints;
- environmental characteristics—geographic location, traffic, soil type, corrosivity, zoning, construction density, groundwater, seismic fault zones, or other geological hazards; and
- indicators of physical condition—breaks and leaks (location, type, and date).

Sources of data can be regrouped in two main categories based on their format, i.e., mainly whether the information can be organized in digital spreadsheets at the pipe level (mostly Excel, Access, or GIS*) or exists primarily as paper records (purchasing reports, index cards, maps, paper listings, paper break reports). The sources may have evolved over time (e.g., paper break reports, then Excel spreadsheet, then GIS). Data sources and format affect the type of analysis that can be conducted. For high-consequence assets such as large-diameter transmission pipes, it is often worthwhile to review the record drawings and specifications and to interview long-term utility staff to ascertain information to help with the assessment.

Data Cleanup

A thorough review of the data is necessary before any analysis because the availability and quality of the data determine the type of analysis that can be undertaken. Data cleanup is often necessary regardless of the type of analysis, whether simple or more advanced (even a simple break rate cannot be correct if the information about breaks and pipes is not well known). However, before data cleanup, it is important to first understand the structural problems in data collection and consider changes to ensure that the quality of future data improves.

If data come in the form of paper records, the level of analysis that can be undertaken will be more limited. In such a case, the main recommendation is to start organizing data for pipes and breaks *at the pipe level in a spreadsheet* or within GIS layers (if the utility has a GIS), paying attention to the types of issues often encountered at utilities that already have their data structured this way. Common problems pertain to data that are

- not available—often the case for environmental factors (traffic, soil, etc.) or abandonment data;
- not linked—break data available but not assigned to pipe;
- difficult to retrieve—data available but costly to retrieve (a cost/benefit analysis must be conducted before engaging in new data collection);
- missing—attribute exists but data field has gaps (e.g., soil information solely at construction or break sites, break data missing in some areas or for some years);
- assumed—data available but has been assumed (material based on the year of installation);
- incorrect—data available but in some cases incorrect (e.g., PVC installed in 1900, break wrongly assigned to new pipe when break actually occurred on original, now abandoned pipe);
- incoherent—break before installation or after abandonment;
- questionable—mismatch between different sources (e.g., break database describes a certain pipe's material as AC, but pipe database describes it as DI);
- not properly defined—leak on joint or third-party break defined as "degradation" break (or vice versa); and
- duplicates—pipe (in some cases due to reusing the same ID when a pipe is replaced) or break (multiple entries or multiple breaks?).

Furthermore, when the sources of data have evolved over time, other problems may arise based on when the data were collected.

* Geographic information system—a spatially organized database. This is the preferred option as it enables better data organization and facilitates analysis.

Cleanup is essential. It can be tedious and has to be carefully planned (some cleanup may not be worthwhile). The more pipe and break data that are in good condition, the better. In some cases, it is preferable to select a few years of good break data (5 years minimum needed) as opposed to more years of data of questionable quality. A preliminary study is required to determine the condition of the data and the cost-efficient cleanup effort. Past problems can be fixed using a combination of the following approaches.

- Manual—consulting maps; design, zoning, or break records; or field personnel
- Logical—running statistical analyses, e.g., if the pipe material is missing, the distribution of material versus location, year of installation, and diameter may help narrow down what the pipe material is most likely to be
- Spatial—using logics along a pipe path or statistics within a buffer zone
- Automatic—algorithms exist or can be developed that can automate steps and save considerable time

Logics and algorithms may be the only options when records simply do not exist. Even if they do, checking records manually can be very labor-intensive and costly, so logics and algorithms may in some cases provide answers that are good enough, at a much-reduced cost.

In the reporting of results, a discussion of data quality and cleanup efforts should be provided to help qualify the validity of the results. For example, the report should indicate what percentage of documentation of the pipes or breaks could not be kept and the reasons these data were discarded.

Breaks and Break Rates

For desktop analysis, generally the most important data are the records of main breaks. When a utility has collected these data for many years, they are invaluable in estimating pipe life expectancy and determining which mains are most in need of attention. AWWA's Utility Benchmarking Program defines a leak as a hole or crack in a transmission or distribution pipeline, valve, hydrant, appurtenance, or service connection that continuously loses water but does not result in a service disruption; and a break as a physical damage to a pipe, valve, hydrant, or other appurtenance that results in a service disruption. Leaks and breaks of lateral services lines are not included. While these definitions are not universal, they are useful for statistical analysis. Both a small leak and large rupture represent a material failure of the pipe, whose causes may be similar even if the severity of the consequences is different.

In most cases, a break does not mark the failure of the main any more than a dead battery marks the failure of an automobile. A typical main is fixed multiple times during its lifetime. "Failure" occurs when the main is deemed no longer serviceable (generally due to excessive repair frequency, etc.) and is replaced.

To compare breaks from one utility to another, the "break rate" is often used. In the United States, this is normally defined as the number of breaks experienced annually per 100 miles of pipe. Break rates between 15 and 30 are fairly normal in the United States, although many utilities set goals that are lower.

Although breaks are expected and tolerated on most water mains, this is not true of all water mains. For large transmission mains and other critical pipes, a single break may be intolerable. Since break data may not exist for these pipes, a desktop assessment often entails comparing the characteristics of these pipes to what is known about their historical performance in other locations. For example, prestressed concrete cylinder pipe made during the late 1960s to mid-1970s is considered particularly susceptible to ruptures, especially pipes made by one particular manufacturer.

Defining the Extent of an Asset

A "pipe" can have many meanings. It can be a few feet long, several miles long, or anything in between. However, for statistical analysis, the pipes in a system need to be somewhat comparable. In most GIS, water mains are divided into many segments. Whenever there is a change in the diameter, the material, or the installation date, it is appropriate to have a separately defined "pipe." Each of these characteristics can have a bearing on the pipe performance. But GIS also tend to segment the pipe at each valve, tees, cross, and bend. This fragmentation creates "pipes" that are often too short for meaningful analysis. For example, the median length of pipe in the GIS of one US utility was approximately 50 ft. Desktop analyses performed on such short segments could lead to poor-quality decisions.

It is therefore helpful to aggregate short pipes into more meaningful groupings that better align with how decisions will be made. Pipe aggregation can be a manual or automated process. Typically, for utilities managing 100 miles or more of pipe, automated processes are more cost-effective. Common aggregation bases include

- the smallest unit of pipe that can be isolated by valves;
- tees and crosses (excluding service lines); and
- groups of continuous pipe with similar characteristics (diameter, installation era, material, etc.).

Similarly, long pipes may need to be divided into multiple segments so they are in line with other pipes being analyzed.

Pipe groupings are not meant to constrain the extent upon which asset-specific decisions must be made. Rather, the intent is to group pipes in a way that more directly aligns with the decisions that will be made.

Defining Pipe Cohorts Through Statistical Analysis

Descriptive statistics of pipes and their breaks can be conducted on pipes with reliable data. If after cleanup data are still missing or are of objectionable quality, they should be removed from the statistical analysis. However, a description of all rejected data should be included in the documentation. The goal of a statistical analysis is to determine differentiating and relevant factors that contribute to the performance degradation of the pipes. The factors typically tested are the pipe characteristics and environmental factors listed in the "Data Sources" section of this chapter.

By comparing failure rates at different ages for pipes with different characteristics, the factors that most influence breaks are discerned. The following relationships are commonly found but are not necessarily universal.

- **Break rates increase with age.** Pipes that are 50 years old will break more often than similar pipes that are 20 years old. Confusion can occur, however, when pipes of different eras are compared. For example, very old (and very thick) pit cast iron pipe will often outperform newer spun cast and ductile iron pipe.
- **Break rates increase with pressure.** Pipes subjected to higher pressure will break more often than similar pipes subjected to lower pressure. Confusion can occur, however, when different pressure classes of pipe are compared.*
- **Break rates for iron, steel, and reinforced concrete pipes will be higher in highly corrosive environments than for similar pipes installed in less corrosive environments.** Confusion can occur, however, when the pipes have different levels of corrosion protection.

* Cyclic pressures accelerate the growth of cracks in brittle materials such as cast iron and polyvinyl chloride (PVC), so higher break rates may be found proximate to pumping stations and other sources of transient pressures.

- **Break rates for asbestos cement (AC) and cast-iron pipes will be higher where ground movement occurs.** These materials have low strain limits, making them particularly sensitive to cyclical shrinkage and swelling, settlement, and any downslope creeping of soils.
- **Break rates decrease with pipe diameter.** Small-diameter pipes will break more often than similar larger-diameter pipes. This is because smaller pipes have thinner walls, making them more vulnerable to corrosion, and are also generally weaker, making them more susceptible to ground movement.
- **Pipes that have broken once are likely to break again.** Whether the pipe is young or old, small or large, subjected to high pressure or low pressure, its history of breaks is often the best predictor of its break likelihood.
- **The intervals between breaks become shorter and shorter.** This is a corollary to the first bullet in this list. If the break rate increases, the interval between breaks decreases. For metallic and AC pipes in particular, break rates generally increase exponentially, meaning the interval between breaks can go from years to months to weeks very quickly.

Factors found to be relevant can be used to define groups of pipes that degrade in a similar fashion (cohorts) on which degradation models can subsequently be based. However, the cohorts should not be too small as this would lead to few or no breaks and no statistical significance. The factors can also be used as covariates in multivariable models.

In many cases, the break rate must be weighted to reflect the data. For example, if 20 years of break data are generally available for the cohort, this may not be the case for all its pipes. Some may have been in service for less than 20 years, installed after the first break of the cohort was observed, or replaced. Appropriate weighting of the break rates allows relevant analysis and comparison. Furthermore, the break rate must be adjusted to reflect data that had to be discarded. If 20 percent of the breaks could not be analyzed because their data were incomplete, but we know that those breaks did in fact occur, ignoring these data would yield a break rate that is only 80 percent of the true rate.

SETTING LEVEL OF SERVICE GOALS

Most water mains can be repaired many times before economics alone dictate their renewal, but frequently repairing a main may produce an undesirable level of service (LOS). The LOS goal is a key parameter for performing a desktop risk assessment for water mains, as it provides a base to gauge pipe breaks over time.

Depending on the size, age, and location of the utility, service levels for distribution main failures vary significantly across the United States. In 2018, AWWA conducted a Utility Benchmarking survey in which Water Distribution System Integrity (as measured by annual combined leaks and breaks per 100 miles) was used as one of the key benchmarking metrics. Table 4-1 shows the results from more than 110 utilities broken into quartiles. In the survey, a *leak* refers to "a hole or crack in a transmission or distribution pipeline, valve, hydrant, appurtenance, or service connection that continuously loses water but does not result in a service disruption." A *break* is defined as "physical damage to a pipe, valve,

Table 4-1 2018 AWWA Benchmarking Survey Distribution System Integrity results (combined leaks and breaks/100 mi/y)

75th Percentile	Median	25th Percentile
7.3	16.4	29

hydrant, or other appurtenance that results in a service disruption." Leaks and breaks of lateral services lines are not included.

Where utilities fall along the spectrum may be influenced by when they started tracking the information and what goals have been set. For example, Denver Water's average combined leak and break rate per 100 miles (which in this case does not include leaks or breaks at hydrants or other appurtenances) from 2000 to 2010 was 12. In 2013, they established a LOS systemwide to not exceed 15 leaks and breaks per 100 miles. This figure was influenced by several factors.

1. Denver Water wants to maintain a LOS that meets or exceeds their customers' expectations.
2. Fifteen leaks and breaks per 100 miles is manageable for existing operations staff to repair.
3. The investment level required to meet this level of service over the next 30 years is sustainable.
4. The target was consistent with a systemwide customer outage metric.

It is important for utilities to use discretion when comparing against benchmarking studies. For one, the data may be skewed by how they are collected and calculated by different utilities. Also, many of the results reflect utilities representing a wide diversity of sizes and regions. What is applicable in a small bedroom community may not be applicable to a large city. The best starting point for LOS performance tracking is for utilities to measure against themselves and determine if current service levels are acceptable and should be maintained over time or if improvements need to be made.

EVALUATING THE LIKELIHOOD OF FAILURE DESKTOP SCORE

There are several methods available for determining the desktop condition score or likelihood of failure (LoF) for water mains ranging from simple to more complex that require advanced statistical models. Transmission mains, for instance, rarely fail and therefore using failure or break data to assign a condition score is usually not possible. This being the case, a weighted-criteria approach applying industry knowledge and experience may be the only desktop analysis option.

Risk Assessment Using Weighted-Criteria Scoring

Weighted LoF scoring for transmission systems is a rudimentary method of pipe prioritization. Many different variables can be examined in determining factors that influence pipe failure. Some examples include

- historical breaks;
- age and era;
- material type, joint type, pressure class, and manufacturer;
- pipe size;
- pressure (static and transient/surge);
- soils (corrosivity and potential for movement); and
- corrosion protection and cathodic protection.

The best indicator to start with is historical pipe breaks if records exist. Any pipeline with previous recorded breaks is more likely to break again. These pipes should receive a high likelihood score. The second variable one might look at is age. Although age by itself is not

Table 4-2 Example criteria and weightings for LoF scoring

Criteria	Weight	Score 1	Score 3	Score 5
Pipe age (install date, year)	30%	>2000	1970–1999	<1969
Soils (corrosivity index)	10%	Low	Medium	High
History of breaks	60%	No failure	One failure	Multiple failures

LoF: likelihood of failure

a good predictor of breaks, breaks increase with age in all pipes. Other factors should be selected (if possible) based on actual performance data (if they exist), or experience from other utilities. Although the opinions of utility staff are valuable, sometimes misconceptions exist regarding what is important. Data-driven decisions are preferred over opinions.

Once the list is compiled for each variable, information can then be assessed by adding each of the scores across the different LoF variables and multiplied by the associated weighting factors to get an overall LoF score for each pipe. An example of this scoring is shown in Table 4-2.

Failure Modeling

If sufficient break data are available (which is more likely to be the case for distribution mains), a regression model that takes into account all relevant variables can be used to assign a likelihood of failure to each water main. This approach eliminates the subjective nature of scoring. Features of the model that are important for the analysis of water main failure include the following.

- **Multivariable.** Multiple factors affect the degradation water main performance. Models should preferably account for these factors simultaneously; otherwise, biases may be introduced.
- **Variable significance.** Often two variables are correlated, but only one contributes to likelihood of failure. Only the significant variable should be used in the model. As an example, in one system, both pressure and ground slope were major factors in pipe breaks, but when the analysis was performed with pressure controlled, ground slope proved to be insignificant.
- **Outputs.** Model output results can take many forms. These include probability of break, probability of failure, time to next break, number of breaks by certain dates in the future, likelihood of failure by a certain date. (Note the distinction, as discussed previously, between a break and a failure.)
- **Breaks on discarded mains.** When a main is discarded, its data are also often discarded; however, this information is helpful in modeling the cohort. Generally, the more data, the better. It is recommended that data for abandoned mains be kept, but the record must clearly keep these data separated from the active replacement main. In the case of partial main replacement, new, old, and abandoned mains will require separate identification.
- **Multiple breaks before failure.** The model should take into account the possibility of multiple break occurrences on one water main before "failure" of the main. Typically, a main is not replaced until it has experienced several breaks. The model should also recognize that a repair does not make the main "good as new."

Output results can be used as LoF scores for each water main—directly if the output is a probability of failure. If the model is forecasting breaks, then a definition of failure needs to be developed to translate breaks to the LoF score. For example, depending on the length and location of and other factors associated with the main, three breaks may be defined as the point at which the main has failed and will be replaced. Estimating the consequence of break ("consequence of failure") is important in determining how many breaks can be tolerated.

Consequence of Failure Desktop Scoring

The pipe consequence of failure (CoF) scoring analysis identifies the impacts that can occur if the main breaks and should be developed to support a risk-based prioritization of either pipe renewal or inspection needs as part of the overall desktop assessment. Industry best practices evaluate CoF in terms of the social, economic, and environmental impacts that the break will cause—typically referred to as a "triple bottom line" analysis approach. Depending on the data available for the pipes, there are three methods that can be used to assign a CoF score, ranging from simple to more advanced. These include

- a weighted-criteria approach,
- a GIS-based approach, and
- a monetized approach.

The weighted-criteria approach is the simplest; specific criteria are selected in each of the triple bottom line categories. The analysis can be performed in a spreadsheet using data obtained from the GIS. An example is shown in Table 4-3.

Using GIS significantly accelerates the assessment if many pipe segments are involved and takes some of the subjective nature out of the assessment process by using specific measurements from queries. The pipe attributes, critical customer connections, and adjacency factors (roadways, water bodies, railways, etc.) are used to create CoF criteria. The criteria and definitions can be built as queries to run in GIS and the final scores stored there or in a spreadsheet. Instead of weighting the criteria, a decision tree can be created that first applies the highest score in any of the three categories to the pipe and then further refines the score based on the number of high-score occurrences. Table 4-4 provides example criteria for GIS-based CoF scoring, and Figure 4-1 depicts the decision tree logic.

A monetized approach for CoF is becoming more popular as utilities look to compare different asset classes, such as pipes and treatment plants, against each other in terms of risk. In this approach, the estimated cost of failure is used as the basis for evaluation instead of a score including both the direct costs (repair) and indirect costs. One challenge of using this method is that it is data-intensive to accurately estimate the full cost of a break and could potentially create public relations concerns regarding how costs are assigned to the social and environmental impacts.

Table 4-3 Example criteria and weightings for CoF scoring

Triple Bottom Line Category	Criteria	Weighting Factor	Low CoF = score 1	Medium CoF = score 2	High CoF = score 3
Financial	Pipe diameter	30%	>12 in.	12–24 in.	>24 in.
Financial and social	Arterial street	20%	No		Yes
Social	Rail- or freeway crossing	10%	No		Yes
Social	Serving a critical customer	30%	No		Yes
Environmental	Crossing water body or environmentally sensitive lands	10%	No		Yes

Table 4-4 Example criteria for GIS-based CoF scoring

Category	Criteria	Evaluation	Initial Ranking
Economic	Diameter	$\leq$8 in.	Low = score 1
		>8 in. and <16 in.	Moderate = score 2
		$\geq$16 in.	High = score 3
	Pressure	<80 psi	Low = score 1
		$\geq$80 and $\leq$120 psi	Moderate = score 2
		>120 psi	High = score 3
	Ground slope	$\leq$15%	Low = score 1
		>15% and $\leq$50%	Moderate = score 2
		>50%	High = score 3

CoF: consequence of failure, GIS: geographic information system

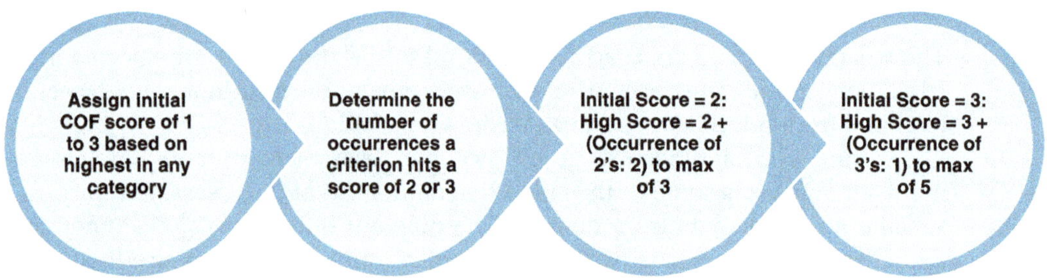

CoF: consequence of failure, GIS: geographic information system

Figure 4-1 Decision tree process for GIS CoF scoring

Risk Scoring

When data are used to compute both the statistical likelihood and cost of a break, the cost of the risk is the mathematic product and can be expressed as a cost per year. This allows a utility to determine whether a risk mitigation (such as relining the main) is worth the expenditure. In making these risk-cost calculations, however, it should be understood that the likelihood of the break and the estimated costs are uncertainties. Direct condition assessment of the mains, as described in other chapters, helps reduce these uncertainties. Reduction in uncertainty has value; it reduces the likelihood of an erroneous decision.

Where data are lacking (or are too uncertain) for computing the risk cost, a risk score is often developed. A risk score helps prioritize inspections for transmission mains or renewal for distributions systems, but these scores are relative, not absolute. Knowing that the pipe on Elm Street has a higher risk score than the pipe on Maple Street does not mean the main will be a problem next year—or even in the next 10 years—but it give the utility an indication of where to focus efforts.

Often, it is helpful to rank the mains on a matrix, as shown in Figure 4-2.

Asset management best practices define the risk equation as

$$\text{RISK} = \text{LoF} \times \text{CoF} \times \text{Redundancy (risk mitigation factor)}$$

Redundancy could also be considered if there are alternate routes to move the water. Typically, the pipe is scored as no redundancy, partial redundancy, or full redundancy. Hydraulic models can help support the redundancy evaluation. For example, the risk score of a pipeline with an LoF of 2.5 out of 5, a CoF of 2.3 out of 5 (both LoF and CoF are obtained using a weighted-criteria approach), and partial redundancy of 0.6 is scored:

$$\text{RISK SCORE} = 2.5 \times 2.3 \times 0.6 = 3.5$$

If, for the above example, the maximum score is 15, the pipeline in question is most likely not a candidate for near-term inspection or renewal unless the CoF or LoF score is at least a 5.

Each utility typically sets a range for its risk thresholds, including maximum CoF and LoF scores to guide different interventions. Lower-score assets are typically kept on a routine preventive maintenance program or can be run to failure; medium-score assets may be on a routine or predictive maintenance program depending on their CoF, and higher-score assets are typically candidates for short-term renewal and or detailed condition assessment. A sample color coded risk map output is shown in Figure 4-3, where red indicates the highest-risk mains, yellow indicates medium risk, and blue indicates low risk.

In addition to a risk score, opportunity criteria can be included in the prioritization process, including

- addressing hydraulic capacity or water quality issues and
- coordinating main renewal projects with road, sewer, or other public works projects.

Project Boundaries

Utilities often group pipes in a more meaningful way to conduct desktop studies and apply effective automated risk score. But in most cases repair, rehabilitation, or replacement projects will not address all pipe segments with a group. Often, it is prudent to adjust the

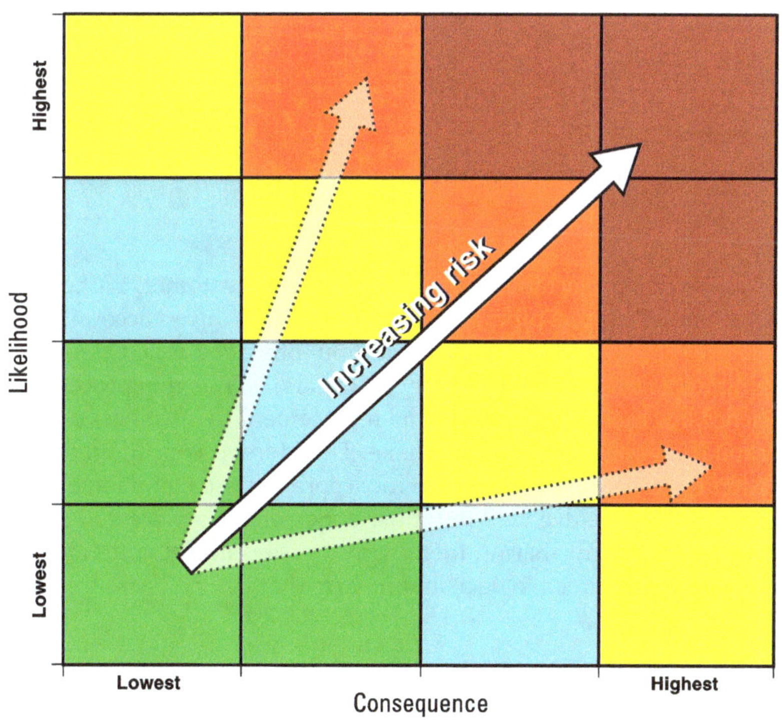

Adapted with permission from Arcadis

Figure 4-2 Risk matrix

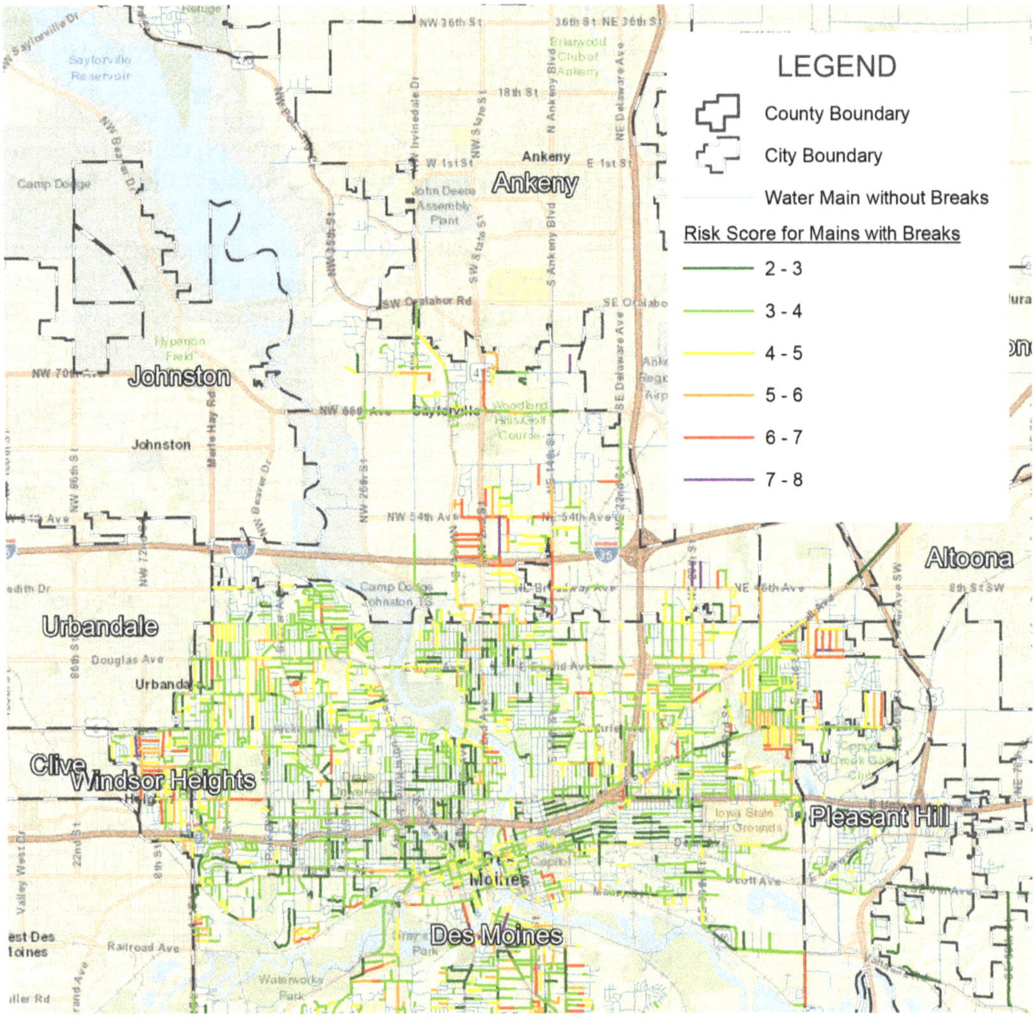

Adapted with permission from HDR

Figure 4-3 Example of color-coded risk mapping

boundaries of a project to limit disruptions (e.g., customer, traffic, etc.); reduce unit costs; replace nearby suspect pipe; account for surface features; account for similar construction techniques being used in the area; and limit other social and political impacts (e.g., avoid digging up a street twice within several years). While this process is as much an art as it is science, for most utilities, pipeline replacement investment constitutes one of the largest capital investments they will make. Therefore, it is prudent to develop guidelines for decision makers regarding how to incorporate these factors in a consistent and effective manner as they are determining the exact extent of a project. Documenting these procedures in writing and sharing them with decision makers often leads to more consistent, transparent, and effective decision making that can be passed on to the next generation of managers.

LONG-TERM RENEWAL PLANNING

Creating Decay Curves (Aging Function)

The aging function shows the break rate based on age. If data availability and quality allow for the type of statistical analysis and modelling described previously, utility-specific aging curves can be drawn using the utility's pipes and break data. Furthermore, if the model predicts breaks per pipe and per year, results can be plotted for each cohort or sub-cohort. The aging curve helps determine the utility-specific effective useful life (EUL) of the cohort.

If data quality and availability do not allow conducting a statistical analysis and modeling, the aging curve can be drawn using the knowledge of its general shape (exponential for most pipes, logarithmic for plastic pipe) and two points to calibrate the curve: one defined by the current break rate and average age (in Figure 4-4, two breaks/100 mi/y and 65 y old), and one with the targeted service level and the industry-assumed EUL drawn from industry manuals (20 breaks/100 mi/y and 95 y old).

Industry-defined EULs are often very different from EULs calculated for a specific material in a specific utility. Generic aging curves or EULs are usually not well suited to making important investment decisions. While published EULs may be a good starting point, the utility is advised to collect and use its own data for meaningful results. Paramount is the collection and effective recording of relevant break data.

Determining Effective Useful Life and Optimizing Renewals

The end of a pipe's useful life depends on a utility's tolerance for breaks. If the aging curve has been drawn as described previously using utility data and the LOS (defined by the acceptable break rate) has been set, the EUL can be read directly on that curve: it is the

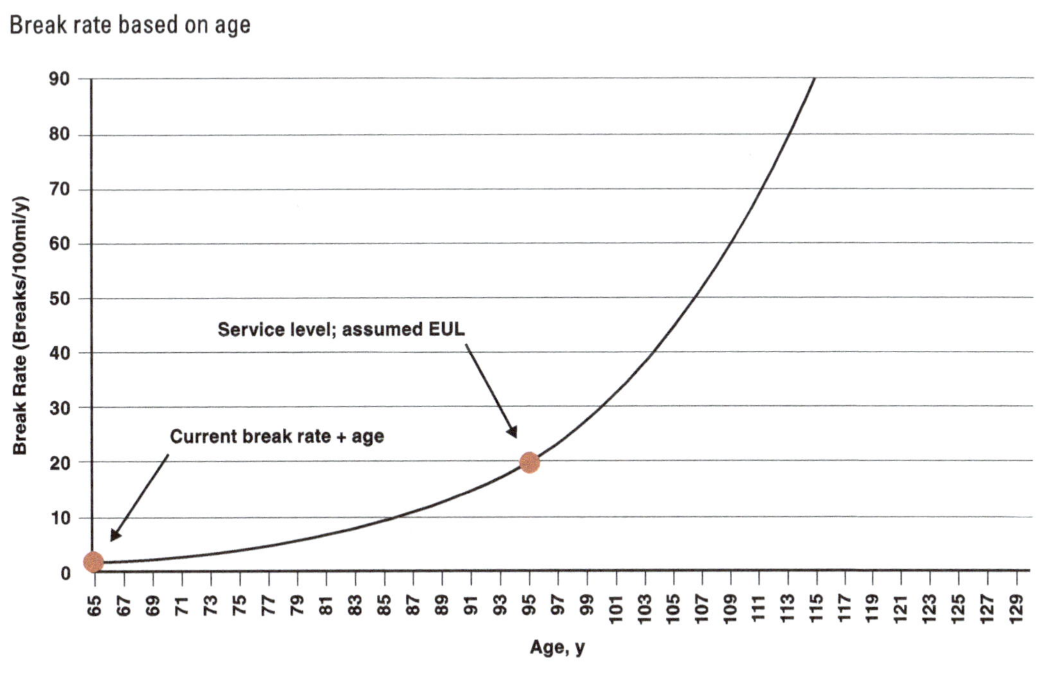

Adapted with permission from Arcadis
EUL: effective useful life

Figure 4-4 Example decay curve based on two known points

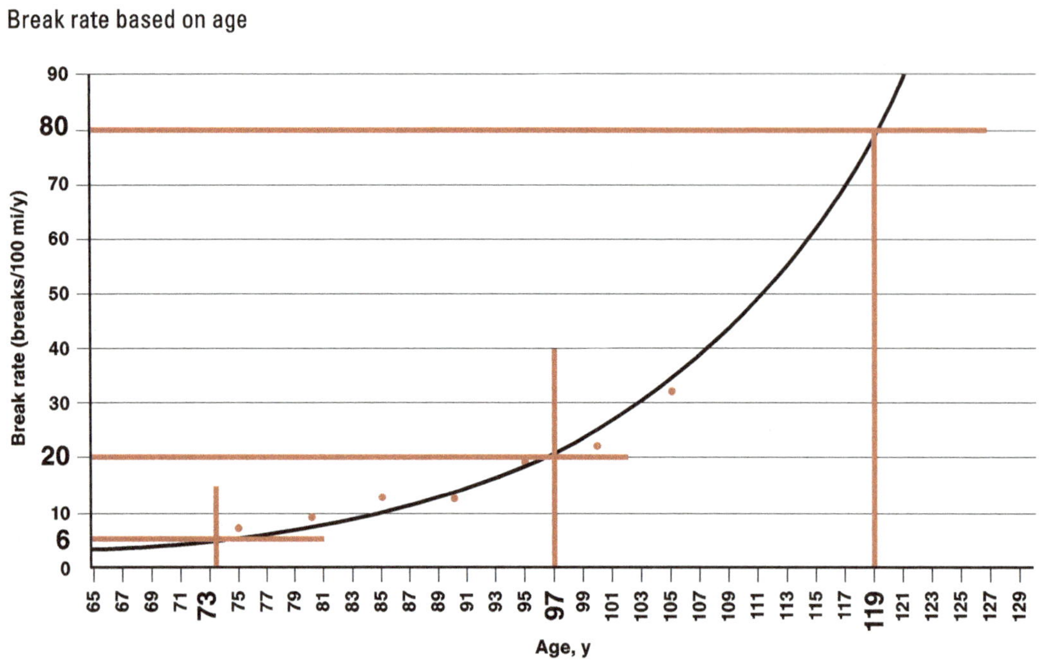

Adapted with permission from Arcadis
CoF: consequence of failure, EUL: effective useful life

Figure 4-5 Example of optimizing pipe EUL by considering CoF scores

age that corresponds to that LOS. Furthermore, the LOS can also be based on the CoF (less tolerance for breaks for a more-consequential pipe), leading to an EUL that is now risk-based. For example, in Figure 4-5, pipes with low criticality are allowed to reach a break rate of 80 breaks/100 mi/y; pipes with medium criticality 20, and pipes with high criticality only 6. This, in the same example, points to EULs of, respectively, 119, 97, and 73 years old. It should be understood that a pipe's break rate can be much higher than the systemwide break rate. In this case, a failure rate of 80 is not inconsistent with a systemwide goal of 15, as long as other mains have break rates of less than 15.

Assigning Replacement Costs

To understand the renewal needs over time, it is important to establish the unit costs for pipe repairs, rehabilitation, and replacement. These costs vary from utility to utility and should be compiled as projects are constructed to form a utility-specific database. Many factors play into the cost per foot for installing pipe, such as depth, pipe size, and pipe location. The costs associated with street pavement restoration is often a major consideration. Engineering, inspection, administration, and permitting costs are also to be considered. When budgeting, appropriate contingency should be included.

Balancing Long-Term Funding Needs and Service Goals

Replacing mains in the short term affects long-term break rates and maintenance costs. Using the techniques introduced in this chapter, it is possible to forecast the number and severity of breaks and to test how these breaks would be affected by various levels of renewal investment. For nontechnical decision makers, it is helpful to communicate this information graphically. Using GIS and spreadsheet tools, a utility can illustrate the effects of service level and risk versus various funding scenarios to select the right plan. The

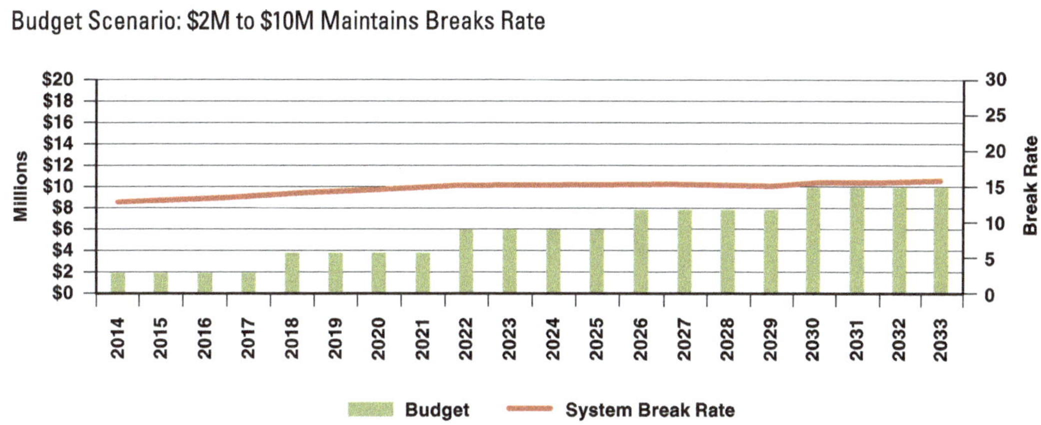

Adapted with permission from Arcadis
GIS: geographic information system

Figure 4-6 Example of using GIS to predict funding needs based on service levels

example output in Figure 4-6 shows how the break rate is affected by a proposed investment scenario.

Data Collection

Desktop analysis relies on good data. Some utilities have collected such data for 100 years, while others have not yet started. With tools that are available today, data gaps can be closed over time. So five and ten years from now, better analysis will be possible. The best repository of such data is a GIS.

The following strategies are helpful when beginning data collection:

- Whenever a main is exposed, document its characteristics: diameter, material, and coating. If a hot-tap is performed, record the lining and wall thickness. Update the records to reflect these data.
- When a main is repaired, in addition to the above, record the date of the break, the type of break (i.e., rust hole, circumferential, longitudinal, etc.) and the exact location (GPS is preferred). Take pictures with metadata (date, location).
- Consider collecting systemwide data for soil corrosivity (conductivity, pH, soil type, and other).
- Obtain readily available maps from the US Geological Survey, Natural Resources Conservation Service, street departments, and others.
- Where a very costly break occurs, or where a series of breaks is unexplained, perform a more detailed analysis, including the collection of samples for possible laboratory analysis.
- Engage staff; seek input from those who repair mains regarding the causes of their failures. Obtain their support in collecting meaningful data.

This page intentionally blank.

Chapter 5

External Corrosion Surveys

Jeff Giddings, Chapter Lead, HDR
David Kroon, Aegion (Corrpro)

Corrosion is the deterioration of a substance or its properties as a result of reaction with the environment. Understanding information about the environment that creates this reaction is key to helping control corrosion, whether the environment is soil, liquid, gas, or some combination thereof. Knowing information about the soil and groundwater (such as resistivity, pH, and chemical composition) informs us of its corrosion potential.

Corrosivity surveys have been used for decades to identify areas where corrosion potentials are high and areas where active corrosion is occurring. By identifying and investigating these areas in greater detail, resources can be targeted where they are most cost-effective. Knowledge of the corrosive environment can be applied to water mains for estimating remaining service life, for setting overall renewal budgets, and for developing more detailed assessment plans.

External corrosion direct assessment (ECDA) is a standard method of using corrosion survey data to select locations for close-up examinations. The pipe is then exposed through excavation and assessed externally using a variety of tools. ECDA is applicable to pipes constructed of iron, steel, copper, and reinforced concrete.

EXTERNAL CORROSION DIRECT ASSESSMENT

In 2002, NACE International developed an international standard for assessing external corrosion of pipelines. This effort was largely driven by the energy pipeline integrity rules and resulted in ANSI/NACE SP0502, "Pipeline External Corrosion Direct Assessment

Methodology." As suggested by the title, this is a methodology that is also applicable to water pipelines. In fact, in a less formalized fashion, the process has been used to assess water line condition for more than 30 years, including iron, steel, and concrete pipe materials.

ECDA is a structured, four-step process consisting of

- pre-assessment,
- indirect inspection,
- direct examination, and
- post-assessment.

The ECDA process incorporates standard techniques for compiling historical information, pipeline and soil surveys, external pipeline inspections, and data analysis. There are no new measurement techniques included in the process. It is the process that is important to successful application of the technique. All four steps must be included.

Step 1: Pre-Assessment

Pre-assessment is the first step in the ECDA process. Chapter 4, "Desktop Condition Assessment," discusses methods for selecting critical pipes meriting direct assessment. Once these pipes are selected, the water utility staff should discuss and agree upon project objectives for the assessment, scope, schedule, and available data. Staff should document pipeline performance and characteristics. All pertinent historical information should be compiled, which may include some or all of the following:

- main and service line composition;
- interconnection of dissimilar pipe materials;
- types of joints;
- electrical continuity;
- location of electrical isolation devices;
- backfill and bedding materials;
- plan and profile drawings;
- construction specifications;
- pipeline lay schedule;
- pipe elevation;
- construction and inspection reports;
- soil report (moisture content, chemistry, and electrical properties);
- groundwater level, fluctuation, and chemistry;
- information regarding adjacent utilities;
- pipeline crossings;
- nearby cathodic protection systems;
- transit systems (particularly using DC electrical power);
- high-voltage overhead power transmission lines;
- cathodic protection data and information; and
- operating history, including repair reports.

When ECDA is determined to be an appropriate assessment methodology, pipe segments should be defined based on original construction contracts or other available documentation. These then form ECDA regions for which indirect inspection techniques are selected for each region. Depending on pipe and environmental conditions, some or all of the techniques described in Step 2 may be employed. In all cases, to comply with the standard, at least two indirect inspection methods should be used. The most appropriate testing technique should be used and then supplemented with a second method of evaluation that complements the first. For example, if soil analysis is selected for one of the techniques, it should be supplemented with a second technique that measures corrosion potential or stray current conditions.

Whichever techniques are selected, the ECDA process requires that they all be performed along the entire length of pipeline in the specified region.

Step 2: Indirect Inspection

Pre-assessment is followed by indirect inspection to identify the locations of aggressive soils and stray current conditions and to define areas where corrosion may be occurring or has occurred. Typical water line survey techniques, which are often used even outside of ECDA, include

- soil resistivity measurements,
- soil sample collection and analysis,
- corrosion potential measurements,
- electrical continuity testing,
- close interval survey, and
- stray current surveys.

All of these survey techniques are well-established. Each has advantages and disadvantages, depending upon the type of corrosion, the pipe material, and the condition of the pipeline alignment. When performing indirect inspections, it is important to

- test the entire ECDA region to cover all points along the pipeline;
- perform all indirect inspections as close to the same time as possible; and
- collect global positioning system (GPS) coordinates for data collection sites and tie data to all permanent features of the pipeline.

Review all available drawings and corrosion data. Perform an end-to-end corrosivity survey of the pipeline, measuring pipe-to-soil potentials, taking Wenner 4-pin resistivity readings, and obtaining soil samples for laboratory analyses (as described later in this chapter). Additionally,

- take photos and notes of evidence of other utilities and general conditions and
- observe rights-of-way access, traffic conditions, and locations and apparent condition of valves, air release valves, test points, and other water line appurtenances.

All data for all surveys should be tied to GPS coordinates to allow for spatial alignment of the various data sets to facilitate data analysis and condition assessment.

Soil resistivity measurements. Soil resistivity is a major determiner of corrosion, and two methods are used to evaluate the soil resistivity: electromagnetic conductivity survey (Emag) and Wenner 4-pin survey (see Figure 5-1).

The advantage of Emag testing is its efficiency because continuous soil resistivity data collection can be performed with a handheld unit that does not require soil contact. It is conducted at accessible locations along the alignment to an approximate depth of 20 ft at 10-foot intervals in accordance with ASTM D6639, "Standard Guide for Using the Frequency Domain Electromagnetic Method for Subsurface Site Characterizations" (2018). Emag testing measures the average soil conductivity and is converted to soil resistivity

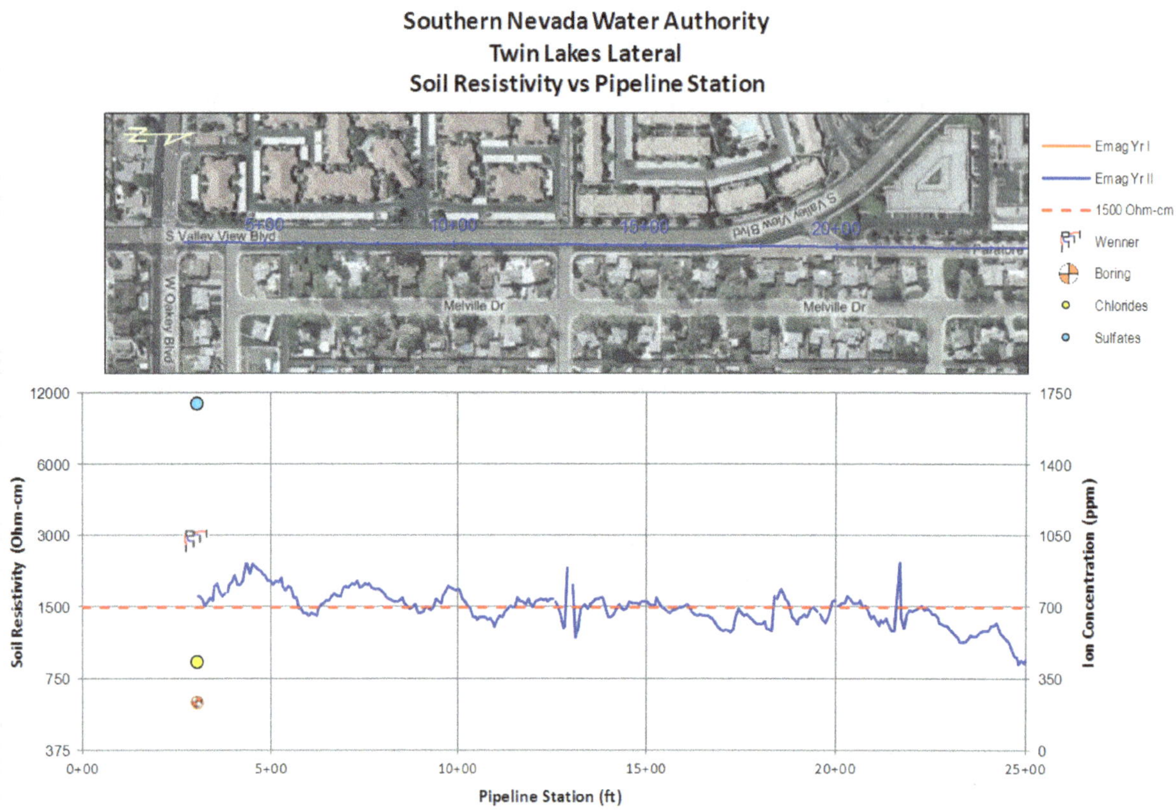

Source: HDR

Figure 5-1 Graphical representation of soil resistivity fluctuations along a pipeline alignment as measured using Emag and comparison to Wenner 4-pin results

(a more traditional unit of measure for determining soil corrosivity) and presented in a graphical format. Because no soil contact is required, this rapid survey method can also be conducted in paved areas, allowing for the evaluation of the entire alignment to determine the most corrosive sections. Thus, areas identified for further study are readily apparent from the graphs. With many miles of proposed and existing alignments surveyed, it is easy to become very familiar with Emag's abilities, as well as its limitations, especially within an existing pipeline alignment where the metallic pipe can affect readings.

Wenner 4-pin testing is used in combination with the Emag survey for further evaluation against industry-accepted categories of corrosivity. Based on interpretation of the Emag survey data, many locations for Wenner 4-pin testing will be selected. In areas where the Emag survey is not conducive to obtaining soil resistivity data (e.g., areas under or paralleling high-voltage lines, congested utility corridors) the frequency of 4-pin measurements should be increased to compensate. Traditionally, the 4-pin tests are completed for nominal depth measurements of 2½, 5, 7½, 10, and 15 ft in accordance with ASTM G57, "Standard Test Method for Field Measurement of Soil Resistivity Using the Wenner Four-Electrode Method" (2012). These measurements are average resistivities measured to the indicated depth. Using average resistivities with the Barnes Layer method, the soil resistivity of each layer can be calculated, which gives a better picture of the soil strata. It is important to measure the soil resistivity in the layers indicated above, because this allows a determination of the actual soil resistivity at pipe depth, eliminating the upper layers (i.e., 0 to 2½ ft and sometimes 2½ to 5 ft strata), which are affected by landscaping or surface vegetation and surface moisture.

Soil sample collection and analysis. Soil samples should be collected at locations selected based on the Emag and Wenner 4-pin data, ensuring that samples are collected at locations representing the lowest soil resistivity areas along the alignment. The samples will be obtained near actual pipe depth using hand-operated equipment, push tools, or soil-boring machines. This task will first require notifying the local utility locating organization and having the underground utilities in the area located and marked.

Soil samples will be analyzed to determine the concentration of soluble chemical species that are known to contribute to corrosion. Typical characteristics to be tested are soil moisture content, conductivity, pH, chloride ion concentration, sulfide ion concentration, and redox potential. The sample analysis results are then used to classify soil corrosivity and estimate corrosion rates.

Corrosion potential measurements. A corrosion potential survey, also known as a pipe-to-soil potential survey, is performed at all accessible test points, including potential, casing, electrical isolation joint, and foreign line test stations, if present, and compared to historical data. Electrical isolation at existing isolation flanges is verified, particularly at main line valves where the pipe material changes, connections to branch lines, and pump stations or other plant facilities. Testing for electrical isolation is accomplished in accordance with NACE International standard SP0286 (2007a). Notes regarding foreign test stations are taken and investigations of potential sources of stray current conducted.

Electrical continuity testing. Electrical continuity testing is important in the overall evaluation of the pipeline because it will indicate whether open joints (i.e., non-welded, rubber gasket, push-on joints without bond cables) will need to be located during the close-interval survey (CIS) and possibly excavated and inspected under the direct examination explained in Step 3. In addition, if the application of cathodic protection is recommended to protect the asset, this information is essential for future cathodic protection design and installation on the pipeline because cathodic protection must be applied to each electrically isolated section of the pipeline.

A test of the state of the electrical continuity along the water line is completed at test station spans to validate the installed condition. Electrical continuity testing is conducted based on pipeline alignment and test station accessibility and completed before conducting the CIS discussed in the next section of this chapter.

Continuity testing is a crucial step and is necessary in order to determine whether the CIS can be conducted with a trailing wire (i.e., small-gage wire connected to a test station via a wire chainer), or if a cell-to-cell CIS survey is required wherein two balanced reference electrodes are progressed along the alignment.

Of the continuity testing methods, the circulating of direct current through the pipe span using adjacent test stations or applicable pipe features is the most accurate method but is time and labor intensive and subject to test station or pipeline accessibility at appropriate distances. This method is excellent for identifying electrically discontinuous pipe joints. In cases where electrical continuity of piping cannot be determined by the circulating current method, as, for example, when insufficient test stations exist to allow for wire spanning between stations, then one or more of the following alternative methods can be used:

1. determining how the pipe reacts to the application of temporary cathodic protection current;
2. measuring earth gradients between two electrodes while applying temporary cathodic protection current;
3. cycling a temporary cathodic protection system and measuring potentials to a fixed reference electrode; or
4. tracing pipe alignment using a pipe locator with a suitably tuned audio frequency.

Close-interval survey (CIS). Following the measurement of potentials at test stations, isolation testing, and electrical continuity testing, a CIS (NACE International 2007b) is conducted along the pipeline based primarily on soil corrosivity, electrical discontinuities, topography, and road hazards. It is expected that pipeline corrosion most likely occurs in corrosive soils and at connections of dissimilar metals, such as those created at changes in water main pipe materials, connections to copper service lines, or main-to–service line connections with brass corporation stops. Before commencing any of the variants of the over-the-line survey, it is first determined if traditional CIS, fixed cell-to–moving cell, or cell-to-cell survey methods will be employed. These test methods are as outlined in Sections A4.5 and A8, respectively, of NACE International standard SP0502, "Pipeline External Corrosion Direct Assessment Methodology" (2010).

The former survey methods will reveal a great deal of information regarding the pipeline, including the levels of corrosion and stray current activity, though only if the data are carefully collected and interpreted. For example, locating the centerline of the pipeline is important in order to collect the correct data. The pipe centerline is determined to within 1 ft using suitable pipe-locating equipment, and depth of cover of the pipeline is determined at the same time and recorded for later use in data interpretation as described below. GPS coordinates must be recorded for all data to allow for accurate data alignment and subsequent interpretation.

When the survey is complete, the CIS, fixed cell–to–moving cell, or cell-to-cell survey data are plotted for analysis (see Figure 5-2). With the exception of the common datum reference, the actual plots of the data are visually similar for the different survey methods

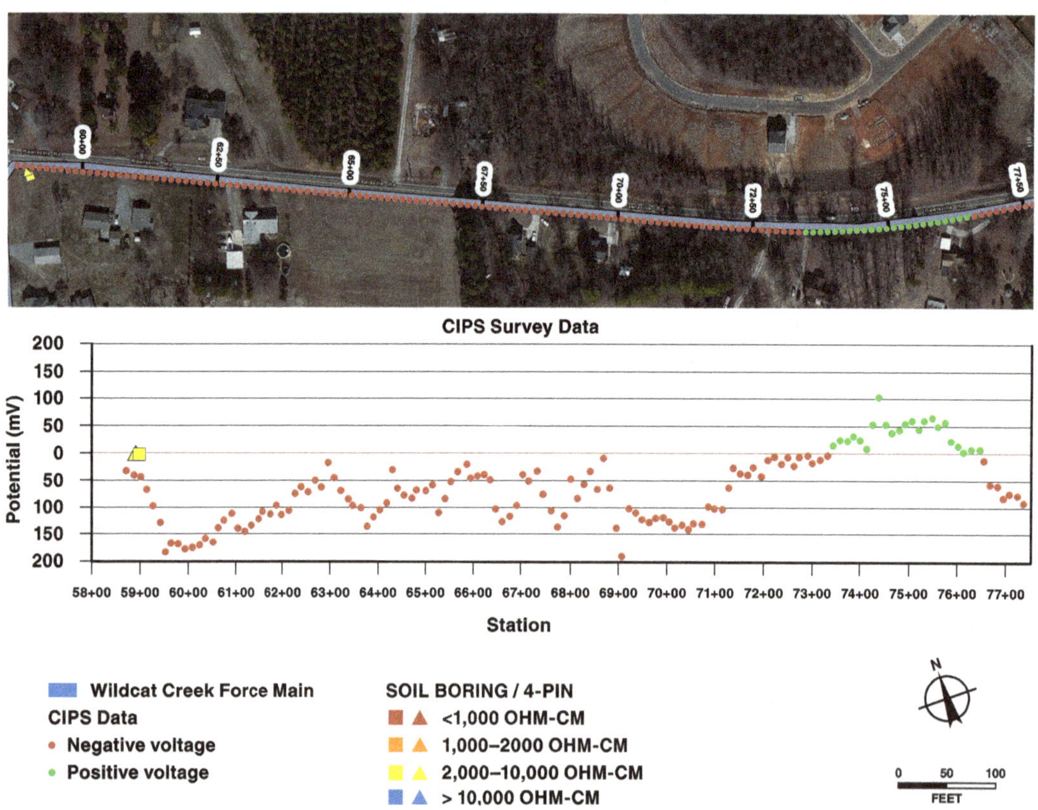

Adapted with permission from HDR
CIPS: close-interval potential survey

Figure 5-2 Graphical representation of cell-to-cell, close-interval survey results along pipeline alignment

and interpretation of the data is the same. The plots are reviewed for significant negative depressions. More negative areas on a pipeline indicate anodic areas where the pipe is corroding to protect more cathodic-adjacent areas.

Data analysis and locations for direct examinations. The raw field data collected from each of the indirect inspections will be reduced to a usable form, tabulated and graphed as appropriate, and analyzed. The pipe depth, soil resistivity, soil sample testing, and CIS data will be used to identify the areas with the highest probability of external corrosion (Figure 5-3). Analysis will include investigating for relevant trends, crosschecking against industry standards, and correlating the data to make engineering judgments. Indications of probable or observed active corrosion from the indirect inspections will be noted and ranked in terms of severity. Collected data and information are analyzed to identify and prioritize assets or locations on the pipeline for direct examination.

Step 3: Direct Examination

The next step in the process is direct examination of the most severe indications to assess the degree of corrosion. This requires excavating the pipe, performing physical inspections, testing the pipe surface, and analyzing the surrounding soil/water electrolytes. Referring to the ranking of indications identified in Step 2, direct examination includes

- excavating the pipe and collecting data where corrosion activity is most likely,
- physically inspecting the pipe,
- collecting soil and groundwater samples,
- measuring corrosion defects, and
- performing a root cause analysis.

Excavation locations to expose the pipe are identified and given priority under Step 2. At least two excavations and direct examinations are required in each ECDA region. Before exposing the pipe, procedures for data collection and record keeping must be defined to ensure consistency. Included should be

- photographic documentation;
- pipe-to-soil potential measurements;
- bimetallic couples;
- surrounding soil, bedding, backfill, and groundwater tests;
- under-film liquid pH;
- coating assessment (dielectric or mortar as applicable);
- mapping and measurement of corrosion defects;
- AC and DC stray current measurements; and
- other analyses, such as microbiologically influenced corrosion (MIC).

Various tools that may be used for spot assessments of the pipe are discussed in Chapter 6. Additionally, methods for physically examining the pipe using "keyhole" tools within small-diameter, vacuum-excavated holes are presented in Chapter 15.

When corrosion defects are found, a root-cause analysis of significant corrosion activity should be performed, and localized corrosion mitigation measures should be implemented on the exposed pipe, such as the connection of sacrificial anode(s) for spot protection.

The process then needs to be evaluated to assess the criteria used to classify the severity of the indications and reclassify locations as appropriate.

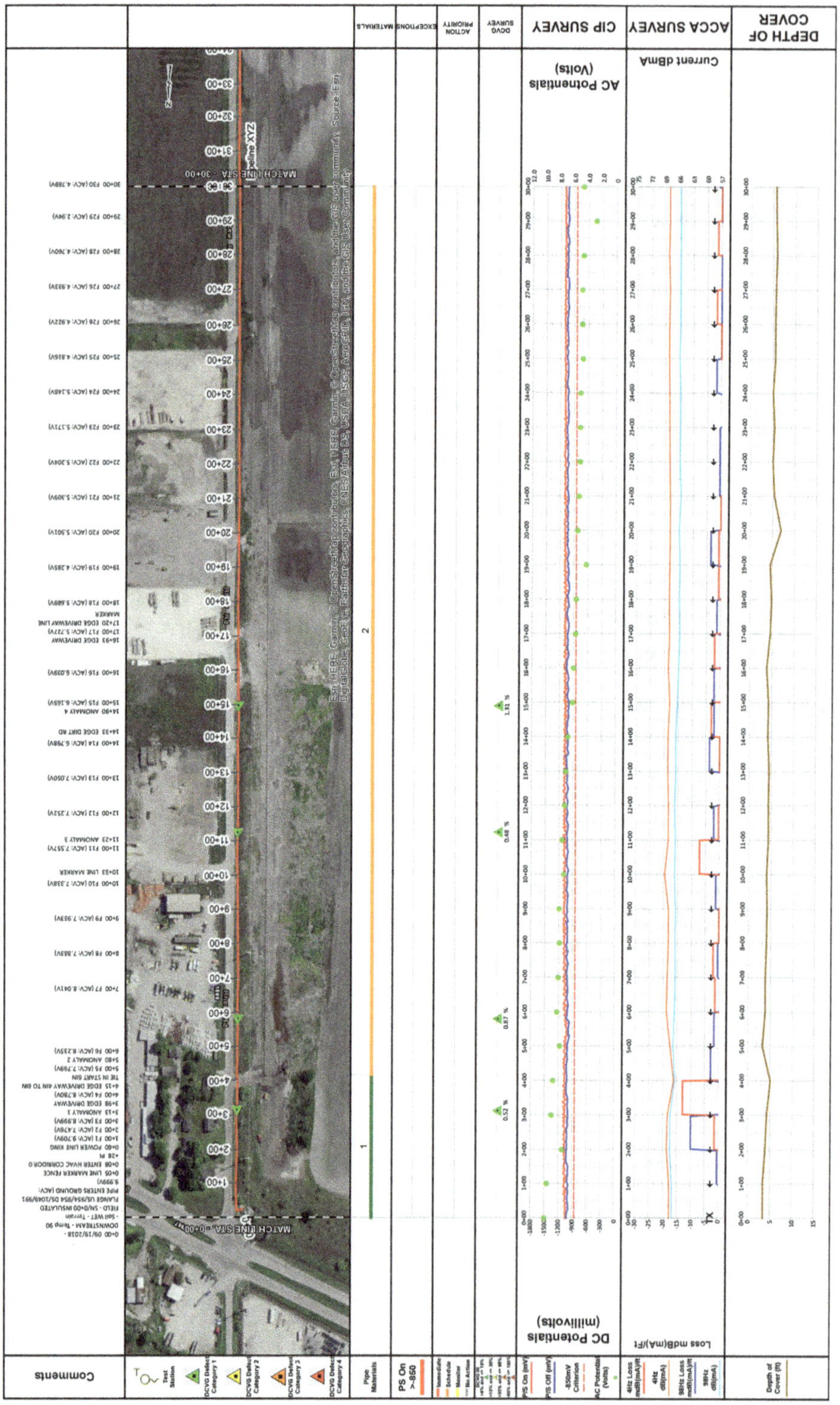

Courtesy of Corrpro, a subsidiary of Aegion Corp.

Figure 5-3 Depth of cover, soil resistivity, and close-interval potential survey indirect inspection data aligned along pipeline rights-of-way

Step 4: Post-Assessment

The last step in the ECDA process is post-assessment to evaluate the overall effectiveness of the assessment and to calculate the remaining life of the pipe.

Various methods can be used to calculate the remaining life of the pipe from the corrosion growth rate and wall thickness as discussed in Chapters 1 and 4. Under some conditions, corrosion rates can be estimated by measurement of linear polarization resistance (LPR) or electrochemical impedance spectroscopy (EIS) tests, as discussed in Chapter 6, through the use of coupons or electrical resistance probes.

The use of ECDA will identify locations where external corrosion activity has occurred, is occurring, or may occur. To aid in assessing the overall effectiveness of the process, an additional direct examination is required where little to no corrosion is anticipated. To maximize the benefits of the ECDA process, ideas for improvements and feedback from people involved should be captured and incorporated in future ECDA.

Although not discussed here, internal corrosion should be evaluated as it is a major risk, especially related to sewer piping. If, during the ECDA process, it is determined that internal corrosion is a threat to the pipeline infrastructure, then a similar process, internal corrosion direct assessment (ICDA), can also be employed. This is particularly relevant for sewer lines where a number of corrosion mechanisms can lead to active corrosion and pipe deterioration. Locations where hydrogen-sulfide gas accumulates in the sewer pipe are particular concerns. Visual assessments (Chapter 8), some leak detection methods (Chapter 7), and acoustic-velocity testing (Chapter 10) can detect these "gas pockets."

ECONOMIC BENEFITS

The greatest economic benefit of managing corrosion comes from improved decisions for pipe repair, rehabilitation, and replacement. System segments are prioritized to target the most important areas first, allowing the best use of capital.

Industry estimates indicate that implementing a cathodic-protection program can reduce the number of water main breaks dramatically. This was demonstrated by large-scale installations in Canada where, after a relatively short transition period, main breaks decreased by 90–95%. Studies of comprehensive corrosion management programs have demonstrated benefit-to-cost ratios from 6.5:1 to 25:1 (Kleiner and Rajani 2002).

REFERENCES

ASTM International. 2018. D6639. Standard Guide for Using the Frequency Domain Electromagnetic Method for Subsurface Site Characterizations. Conshohocken, Pa.: ASTM International.

ASTM International. 2012. G57. Standard Test Method for Field Measurement of Soil Resistivity Using the Wenner Four-Electrode Method. Conshohocken, Pa.: ASTM International.

Kleiner, Y. and Rajani, B. 2002. Quantifying the Effectiveness of Cathodic Protection in Water Mains. National Research Council of Canada, Institute for Research in Construction, presented at 2002 NACE International Conference, Montreal.

NACE International. 2010. SP0502. Pipeline External Corrosion Direct Assessment Methodology. Houston: NACE International.

NACE International. 2007a. SP0286. Electrical Isolation of Cathodically Protected Pipelines. Houston: NACE International.

NACE International. 2007b. SP0207. Performing Close Interval Potential Surveys and DC Surface Potential Gradient Surveys on Buried or Submerged Metallic Pipelines. Houston: NACE International.

This page intentionally blank.

Chapter 6

Spot Assessment

Mersedeh Akhoondan, PhD, PE, Chapter Lead, HDR
Sylvia Hall, PE, Sylvia Hall Engineering

This chapter provides an overview of available techniques for external spot assessment of linear assets. The topics addressed in this chapter should provide insights regarding the following issues:

- selecting locations for spot assessment,
- strategies that enhance the accuracy of spot assessment,
- in situ field tests commonly used for spot assessment,
- laboratory tests commonly used for spot assessment,
- interpreting the acquired data, and
- achieving a balance between data resolution and feasibility.

As an old proverb says, there are no better witnesses than your own eyes. Direct assessment techniques are invaluable for determining the condition of buried structures.

In an ideal case, agencies have enough time, budget, and resources to expose the pipe and conduct direct assessment of their entire portfolio of aging assets (i.e., miles of pipelines). In practice, such extensive assessments seldom occur and are not economically justified. When used properly, spot assessment is an economical method that can provide valuable insight about the condition of a buried asset.

Spot assessments are preferably performed at areas of elevated corrosion concern, as determined by over-the-line surveys such as those discussed in Chapter 5. In other cases, spot assessments are performed in areas where indirect assessment tools are ineffective or impeded. Examples include areas where coatings cause electrical shielding and interfere with indirect assessment tools' signals or where backfill materials, frozen ground, or neighboring underground structures interfere with the function of other assessment techniques and limit the adequacy of indirect testing. Spot assessments may also be performed on pipelines that have been indirectly assessed for the first time in order to verify and calibrate the result of other assessments. This is more common for pipelines that do not have a good record of corrosion protection. Spot assessment may also be performed to verify other assessment results. While the focus of this chapter is on targeted or planned spot assessments, the same techniques could be applied for opportunistic spot assessment; for example, saving and examining tapping coupons upon removal from the tapped main provide an opportunity to observe the conditions of the water mains internally and externally.

The next few sections discuss the best approaches for effective spot assessments of water mains. Approaches discussed here may not apply to every specific situation and complexity. The intent of this chapter is to serve as a guideline and not as a road map. Spot assessments should occur after pre-assessment techniques (i.e., desktop analysis of historic data evaluation, indirect assessments, and over-the-line corrosion surveys) have been performed.

STRATEGIES FOR EFFECTIVE SPOT ASSESSMENT

Selection of Locations for Spot Assessments

Excavation and exposure of the buried asset can be costly. The average direct cost of a single excavation is estimated at $3,000 and, depending on the location, depth of the pipe, traffic conditions, and surface restoration, can be considerably more. The indirect social costs (e.g., effects to neighboring businesses and impacts on traffic) make it important to select the right location(s) for spot assessment to acquire the optimum information and reduce the need for additional costly excavations. Two common approaches typically considered when selecting the locations for external spot assessment are as follows.

- If no information pertaining to the condition of the pipeline exists, conduct feasible over-the-line surveys (such as soil corrosivity and potential surveys discussed in Chapter 5) and acoustic velocity techniques (discussed in Chapter 10) to identify the locations with the highest risk of corrosion. Carefully evaluate and make an engineering judgment about the criticality of each location. While sophisticated risk management tools are available to rank the risk associated with various assets, a simple approach is to locate the areas with high consequence of failure (e.g., critical customers such as hospitals, schools, military bases). See Chapter 4 for more on desktop risk analysis. Based on site accessibility, select the locations with the highest likelihood of active corrosion and the greatest consequences of failure.
- If some information pertaining to the condition of the pipeline exists, select the locations where indirect inspection techniques were precluded or the results were uncertain. Also, if indirect assessment techniques were used for the first time on the pipeline, perform spot assessments at a few tested areas to verify the results. In this case, select the areas with high data resolution and confidence intervals.

System operators often have institutional knowledge and previous failure/break information that should also be considered when identifying the locations of active corrosion for potholing. Joints can sometimes be the most problematic areas. Leakage at joints due to misalignment, gasket/joint assembly, material degradation, and poor quality of field-applied mortar are common. Impressed-current cathodic protection can cause damage and corrosion at pipe joints if the pipeline is not electrically continuous or if bonds have broken. Therefore, it is a common practice to expose pipe at joints, particularly if the electrical continuity (external bonding) is to be tested or established. Locations of new valve installation or replacement are also good candidates for spot assessment.

To develop baseline information, spot assessments are also performed on locations where no degradation or corrosion is anticipated. In this case, two locations along comparable pipe material (similar materials, age, fabrication, and exposure), one at a spot where active corrosion is anticipated and one where no corrosion is anticipated, are chosen for spot assessments. Comparing the two conditions helps to understand the mechanism of degradation and to determine the effective method of future corrosion protection.

Before excavation for spot assessment, the location of the pipe should be carefully determined based on as-built drawings and geo-referencing tools. If the exact location is unknown, above-ground pipe-locating techniques can be used to determine the pipeline alignment. Such techniques typically involve placing a transmitter on the pipe and using a receiver along the suspected pipeline alignment. Based on the strength of the received signals, the location of the pipeline is determined.

Opportunity assessments. Because excavations are costly, it is important to take advantage of excavations performed for pipe repairs, service connections, and other projects. Repair of leaks and breaks, in particular, can provide valuable information about the condition of the pipe.

However, any single spot assessment is difficult to interpret without additional information about the pipe, its repair history, and its environment. An assessment performed at a repair may reveal badly deteriorated pipe that may not be indicative of most of the pipeline. And an assessment performed at a service tap (or at a random location) is likely to reveal pipe that is in relatively good condition, providing a false sense of security; a break may be pending a few feet away. Thus, spot assessments may need to be conducted at several locations or combined with data from other assessments to provide a good picture of the condition of the pipeline.

Utility Clearance and Permitting Before Excavation

Once the excavation locations are selected, utility clearance is needed before excavation. This is often done by contacting a utility clearinghouse (such as Underground Service Alert for the State of California). If the clearinghouse is unknown, then calling 811 from anywhere in the United States will automatically be routed to one call center. Similar services are available in other countries. After providing the information about the location and the purpose of the excavation, each affected utility company with potential facilities at the site will generally deflect the sound waves, and no reading will be obtained is notified and must respond within a specified time frame (typically 36–48 hours) to mark the approximate location of its underground utility lines. Per local or state regulations, a clearance distance (typically 2–3 ft) is needed to avoid damaging utilities lines when performing spot assessment.

If the utility markings indicate that interferences are present, a different location may be needed for the excavation. Selecting a few alternatives for spot assessment to account for unforeseen situations that prohibit excavation at particular locations, such as utility crossings, site inaccessibility, and permit issues, is recommended.

The amount of excavation needed for spot assessment depends on the size of the pipe, the extent of corrosion activity, and the assessment techniques that will be employed. It is generally recommended to expose the entire circumference of the pipe and extend the excavation to 1 to 1½ ft below the invert to provide enough space for visual inspection and testing. For very large pipe where full excavation around the circumference of the pipe is very costly, the excavation may be reduced to at least the 12 o'clock to 6 o'clock position on one side to provide access to the top and bottom of the pipe for external assessment. As described in Chapter 15, assessments using small, vacuum-excavated "keyhole" excavations may be appropriate.

Before excavation in a public street, a permit will generally be required. For busy streets (or for any street in some jurisdictions), this may require the development of a traffic control plan. If the assessment requires that a person enter the excavation, appropriate shoring and other worker protection may be needed. Shoring may also be needed where the excavation is deep or where soils are unstable.

Precautions

Every effort should be made to preserve the condition of the pipe during the excavation before spot assessment. Excessive soil movement and surcharge loads due to mobilization of heavy equipment may introduce significant stress to certain pipe materials (e.g., asbestos cement pipe) and inadvertently worsen the prevailing pipe condition. Direct physical and mechanical damage to the pipe (e.g., by shoveling) should be avoided. Photo documentation of as-received condition immediately after exposure of pipe should be done to set apart environmental effects (if any) of exposure such as flash rusting.

Excavation and working around pressurized pipe could be dangerous, and all safety precautions specified by government regulations and industry standards must be followed.

TESTING PROCEDURES

This section provides a summary of common in situ field and laboratory testing for spot assessment (Table 6-1) that can be tailored to specific pipeline situations.

In situ Field Testing

Before excavation, the type of tests, the minimum requirement for consistent data collection, and record-keeping requirements should be specified. Such requirements should

Table 6-1 Common in situ field and laboratory tests for spot assessment

Test	Applicability
Visual inspection	All types of pipe
In situ pH testing	Concrete, cement-mortar lined/coated, asbestos cement
In situ coating assessment	Coated metal pipe
In situ chloride content	Concrete, cement-mortar lined/coated
Wall thickness and pit-depth measurements	Metal pipe
In situ potential mapping	Concrete, cement-mortar lined/coated
In situ electrical continuity testing	Concrete and metal pipe
Rebound testing	Concrete, cement-mortar lined/coated
In situ electrochemical testing	Concrete, cement-mortar lined/coated
X-ray fluorescence	Metal pipe
Microbiologically influenced corrosion (MIC) testing	All types of pipe
Chemical analysis	All types of pipe and coatings
Elemental analysis (e.g., energy-dispersive spectrometry [EDS])	All types of pipe and coatings
Microscopic analysis	All types of pipe and coatings
Metallographic analysis	Metal pipe
Simulative analysis	Metal pipe based on soil information
Accelerated electrochemical testing	Metal pipe

be based on the pipe materials, types of corrosion activity expected, and conditions to be encountered. The services of a certified corrosion engineer or specialist may be worthwhile.

Visual inspection. A visual inspection will provide crucial information for completing and documenting the results of the corrosion assessment. The key for good visual inspection is to be consistent in the grading and evaluation of what is observed in order to provide a means of comparison. Visual inspection requires adequate illumination of the surfaces and proper eyesight of the tester. To be most effective, visual inspection requires knowledge of pipe materials, anticipated forms of corrosion, service/operational conditions, and acceptance criteria.

To view the metal cylinder, a thorough cleaning is required to remove soil, scale, and corrosion from the pipe. In the case of cast iron and ductile iron, graphite must also be removed to reveal the true extent of pitting and metal loss. Grit blasting is frequently needed.

Photos. Visual inspection should be accompanied by good photographic documentation. "A picture is worth a thousand words," but this is only true for pictures with proper scale, proper focus, good lighting, and detailed identification of locations and features observed.

Coating inspection. When the condition of coatings is being assessed, features to be documented include loose and removable powder at the surface (referred to as chalking); obvious pinholes and cuts (referred to as holidays); coating disbondment; cracking; rough spots or flaking; presence of thin spots, gouges, or chipped areas; and presence of discoloration and rust staining.

Concrete/mortar inspection. For concrete- or mortar-coated steel pipelines, features to be documented include size, location, and orientation of cracks; spalling; exposed reinforcement; delamination; leakage; joint misalignment; and presence of biological organism (e.g., slime).

Metallic pipe inspection. For ductile-iron, cast-iron, and other metallic pipelines, the presence of cracks; the extent and forms of corrosion (localized versus uniform); the color of corrosion deposits; the sizes and locations of pinholes; leakage; the condition of welded areas; and any observed deformations, graphitization, and mechanical damage (i.e., dents, tab holes) should be documented. Pit depth gauges (Figure 6-1) are used to measure the depths of individual pits in relation to the surrounding surface.

Other pipes. For other pipe materials, the presence of cracks, the condition of joints, leakage, and any observed deformation or discoloration should be documented.

Soils. The appearance of surrounding soil/backfill and the presence of groundwater (if encountered) should also be noted and documented.

In situ pH testing. A common method of determining the pH range of cement-based materials is by spraying a solution of 1% phenolphthalein (a pH indicator) in 50–70% ethyl alcohol on freshly exposed concrete/mortar surfaces at various depths (⅛-in. intervals). Phenolphthalein is colorless below pH 8.0 and turns fuchsia pink above pH 10.0. Sound, uncarbonated concrete/mortar is highly alkaline (it has a pH above 12) and will turn pink after being sprayed with phenolphthalein. When using this method for external condition assessment of concrete or cement mortar pipe, the staining should be done at the areas of concern (such as where spalls and rust stains are observed). If the sprayed area does not immediately turn pink (see Figure 6-2), the test may continue on deeper surfaces (by carefully chipping a small area or drilling) and reapplying the phenolphthalein solution. If the test is to be done in a drilled hole, the dust is first removed from the hole. The depth of the uncolored layer (the carbonated layer) from the external surface is measured to the nearest millimeter. Chipping or drilling concrete/mortar should only be done on nonpressurized, non-asbestos cement pipe and only if damage to the coating is acceptable. Chipped areas should be repaired with corrosion-resistant grout or mortar. For pressurized pipes and asbestos cement (AC) pipes, coupons for testing may be collected through proper tapping and cutting procedures.

Source: HDR

Figure 6-1 Measuring the depth of corrosion pits in cast-iron pipe

Source: HDR

Figure 6-2 In situ pH testing of concrete pipe using phenolphthalein

Phenolphthalein stain testing is commonly used on samples of AC pipe to determine the depth of internal and external degradation. Samples are preferably complete "ring" sections (i.e., a full 360-degree section of pipe) because significant variability in degradation depth often occurs around the circumference, both internally and externally. Before performing any work on AC pipe, special procedures and training are needed to ensure that fibers are not inhaled or do not contaminate the area and that materials are properly disposed of. (Stain testing and other spot assessments of AC pipe are discussed in greater detail later in this chapter.)

In situ coating assessment. Various test methods such as dry film thickness (DFT) (ASTM D7091 [2013] and D4138 [2017]); holiday detection (ASTM D5162 [2015]); and adhesion testing (ASTM D3359 [2017], D4541 [2017], D6677 [2018]) can be used to determine the condition of pipe coating (particularly a dielectric coating) as part of spot assessment. These tests should be conducted by trained individuals, and it is good practice to perform the tests in several locations, including the top, side, and bottom of a pipe; the edge of fittings; and areas where wear or damage may occur.

In situ chloride content. Chloride is detrimental to ferrous and many other metals. In sufficient concentrations, chloride ions are also the primary cause of corrosion initiation of steel in reinforced concrete structures. Chloride ions can destabilize and impair the protective passive film on the surface of ferrous metals and consequently cause localized corrosion. The concentration of chloride required at the surface of metal for corrosion initiation is referred to as the chloride threshold. For reinforced concrete, a common practice is to assume a maximum allowable chloride content (per ASTM C1152 [2012]) of about 0.4% by weight of cement.

Test kits for onsite chloride content testing of concrete or mortar lining are commercially available. A sample of concrete or mortar powder is mixed into a measured amount of extraction liquid and shaken for five minutes. The extraction liquid removes disturbing ions (i.e., sulfide ions) and extracts the chloride ions in the sample. A calibrated electrode is submerged into the solution to determine the amount of chloride expressed as percentage of concrete mass. While onsite testing results are likely not as accurate as laboratory results, the test provides some insight on the risk of corrosion by chloride attack and corrosion mitigation plans to be used.

Wall thickness and pit-depth measurements. For metallic pipe, corrosion affects the remaining wall thickness and strength of a pipe; therefore, thickness measurements particularly at pitted areas (localized corrosion areas) are part of routine corrosion inspections. Various handheld pit gauges (e.g., dial-type, contour gauge) can be used to capture the pit depth and profile. Such tools require cleaning all deposits and residues from inside the pitted areas so that accurate measurements can be taken.

A more advanced technique in determining the depth of degradation and pipe wall thickness is to use digital ultrasonic gauges. An easy-to-use handheld digital gauge converts a pulse of electrical energy into a high-frequency ultrasonic sound wave, which travels through the pipe material and reflects off the back wall and returns to the transducer (probe). The thickness of the material where the probe is located is the value that is displayed on the meter. Ultrasonic gauges should be periodically calibrated. The tested surfaces should be clean and free of dirt, oil, grease, paint, and contaminants. In most cases, a couplant (e.g., conductive gels, glycerin) is needed to ensure adequate acoustic contact between the proper probe head and the surface that is being measured. Couplants should be used in accordance with the equipment manufacturer specifications.

Although the use of ultrasonic gauges is relatively easy on steel and other homogeneous materials, difficulties arise in obtaining accurate readings on cast iron. Graphite flakes within the iron will generally deflect the sound waves, and no reading will be obtained.

The data collected may be used to determine the remaining strength of the pipe based on operational pressure (condition). When corrosion is observed, the strength of the pipe must be evaluated to determine if the pipe is still strong enough to contain its internal pressure. Some guidelines (e.g., ASME B31G [2012] for steel pipes and WRF 4233 [2012] for prestressed concrete cylinder pipes) are available to make such determinations. It is easier to perform detailed pit-depth and thickness measurement of the field coupons in the laboratory. As such, this test method can be considered a laboratory test. Of course, the structural integrity of the pipe should always be reevaluated if the condition of the pipe has changed from designed conditions (i.e., operational pressure or loading conditions).

Currently, ultrasonic-based inspection techniques (Figure 6-3) are being developed to acquire information pertaining to defects in buried concrete and other cement-based pipe materials.

Several electromagnetic tools can also be used in spot assessment of metal pipes, as discussed in Chapter 11.

In situ potential mapping. A common test method for identifying the hot spots (areas of active corrosion) for concrete pipe or cement-mortar coated steel pipe is to conduct half-cell potential mapping on the pipe surface per ASTM C876 (2015). Potential measurements are collected in a closely spaced grid pattern (Figure 6-4). From those measurements, equipotential contour maps are plotted and used to identify high- versus low-corrosion activity areas.

In situ electrical continuity testing. Electrical continuity across pipe joints is necessary if cathodic protection is deemed an appropriate mitigation method. Electrical continuity tests, such as the circulating current test, may be performed as part of spot assessments. If pipes are found to be discontinuous, it may be advantageous to bond the joints at the already-excavated site. For large-diameter pipes, bonding can also be done internally.

Source: HDR

Figure 6-3 Ultrasonic thickness measurement

Source: Mersedeh Akhoondan, HDR

Figure 6-4 Potential mapping

Aural testing of concrete. Aural testing using an appropriately weighted hammer (known as "sounding") is a simple way of quickly determining whether areas of concrete/mortar delamination are present. This survey provides an expedient, overall assessment of the structure and will guide much of the remaining testing as to worst-case locations (i.e., those areas where corrosion is most likely to be found).

Rebound testing. Rebound testing using a rebound hammer test (also known as Schmidt Hammer test per ASTM C805 [2018]) can be used on concrete pipe to assess the uniformity and relative strength of concrete by measuring its surface hardness. The measurements (the rebound height of a spring-loaded mass against the surface) are typically taken quickly, but multiple readings are required to obtain accurate and representative values.

In situ electrochemical testing. The voltage response of a corroding metal is different from that of noncorroding metal. The voltage of noncorroding metal increases in the low-frequency zone of the plot, but it is almost invariable for corroding metal. This basic concept has been used in various in situ corrosion rate–monitoring technologies for the detection of the corroding areas. Many devices come with user-friendly manuals and can provide instantaneous estimates of corrosion rates and corrosion potentials. While there might be some uncertainties about the magnitude of obtained corrosion rates (i.e., uncertainties regarding original thicknesses), the measurements at various locations help to identify active corrosion spots. Figure 6-5 shows an example of in situ electrochemical testing.

While the detailed explanation of electrochemical tests (such as ASTM G59 [2014], ASTM G3 [2014], ASTM G5 [2014], ASTM G102 [2015], ASTM STP1188 [1993]) is beyond the scope of this manual, brief descriptions of two common test methods (electrochemical impedance spectroscopy [EIS] and linear polarization resistance [LPR]) are provided. Both of these nondestructive techniques provide estimates of corrosion rates based on the

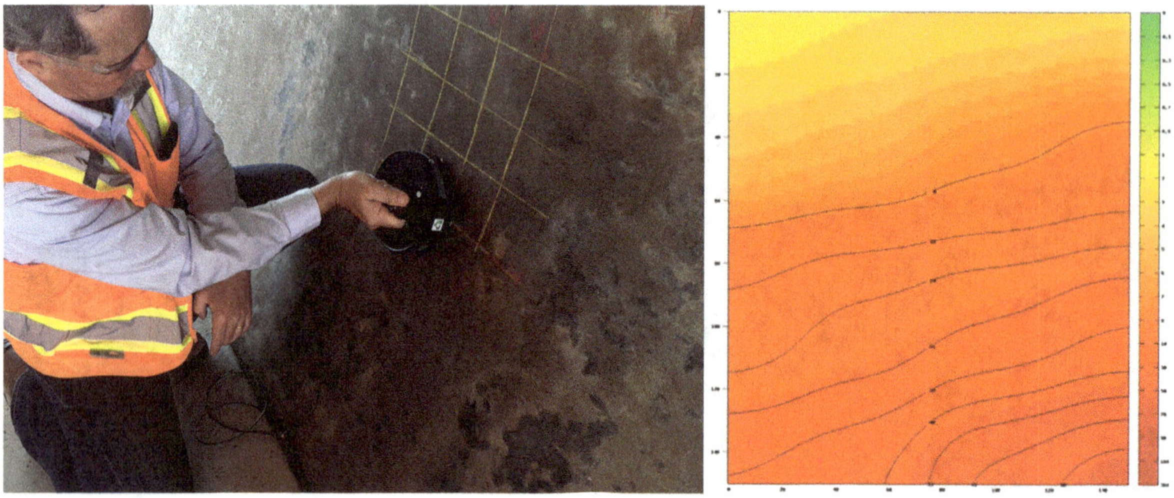

Sources: Mersedeh Akhoondan, HDR, and Giatec Scientific, Inc.

Figure 6-5 In situ electrochemical testing

electrochemistry of the materials. These test methods can be used in the field or in the laboratory.

- The EIS technique involves the application of a time-varying voltage and measurement of the current response. The ratio of the two gives the frequency-dependent impedance. The low-frequency end of the impedance magnitude is typically related to its polarization resistance (a parameter proportional to the inverse of corrosion). This allows the Stearn-Geary equation to be used to estimate the corrosion rates.
- LPR is a common DC technique for measuring polarization resistance. In this method, the potential is scanned about ±5 to 10 mV relative to the corrosion potential. The change in potential with respect to the change in current is calculated to obtain the polarization resistance, which is inversely proportional to the corrosion rate of the metal tested.

In situ elemental analysis of metal pipe using x-ray fluorescence. X-ray fluorescence (XRF) is a nondestructive analytical technique used to determine the elemental composition of materials using a simple handheld tool. The elemental chemistry of a sample is determined by measuring the fluorescent (or secondary) X-ray emitted from a sample after excitation by a primary X-ray source. XRF handheld tools are relatively inexpensive to rent and can be used to verify the composition of pipe and coating materials in situations where uncertainties exist.

Microbiologically influenced testing. Microbiologically influenced corrosion (MIC) refers to the indirect and direct effects that a biological agent has on the corrosion of metal and metal alloys. In this case, biological organisms can alter the corrosion reaction kinetics at the metal surface to either enhance or inhibit corrosion. These organisms may adhere to the metal surface, creating a biofilm layer that can significantly change the characteristics of the metal/solution interface and the corrosion performance of the metal alloy. The presence of tubercles, slime, black iron sulfide deposits, algal growth, or "swamp gas" odors are good indications of possible MIC. A drop of hydrochloric acid on a black sulfide deposit releases hydrogen sulfide and yields a rotten egg odor.

The types of bacteria present in a given sample can be identified by culturing samples in growth media appropriate to the type(s) of bacteria of interest. Commercial kits for field sampling bacteria involved in MIC are available from several manufacturers. The sample tested may consist of water, sludge, slime, or corrosion products.

The test kits vary in approach and complexity and in whether the results are qualitative, semi-quantitative, or quantitative. Some kits test for only one or two types of bacteria. The most comprehensive kits provide quantitative results and test for the five types of bacteria listed in the next paragraph. The kits come with a sealed swab and sealed tubes containing culture media. Small amounts of the substances (suspected to contain bacteria) are collected with the provided swap and placed into the culture tube. The tubes are then kept at room temperature for a few days and, depending on the discoloration of the culture, the amount of bacteria present is determined. The detection range for kits using serial dilution are generally 1 to >100,000 bacteria of each type, although additional tubes can be used to extend the detection range. An example of sampling slime for MIC testing is depicted in Figure 6-6.

Specialized kits may be purchased to test for one or more of the following types of bacteria:

- sulfate-reducing bacteria,
- sulfur-oxidizing bacteria,
- iron-oxidizing bacteria,
- acid-producing bacteria (APB) and fungi,
- slime-formers, and
- low-nutrient bacteria.

All kits come with detailed instructions. Some kits can be returned to the manufacturer for interpretation. The average price of the test kits is around $300, and the instructions and interpretation are typically easy to understand and follow.

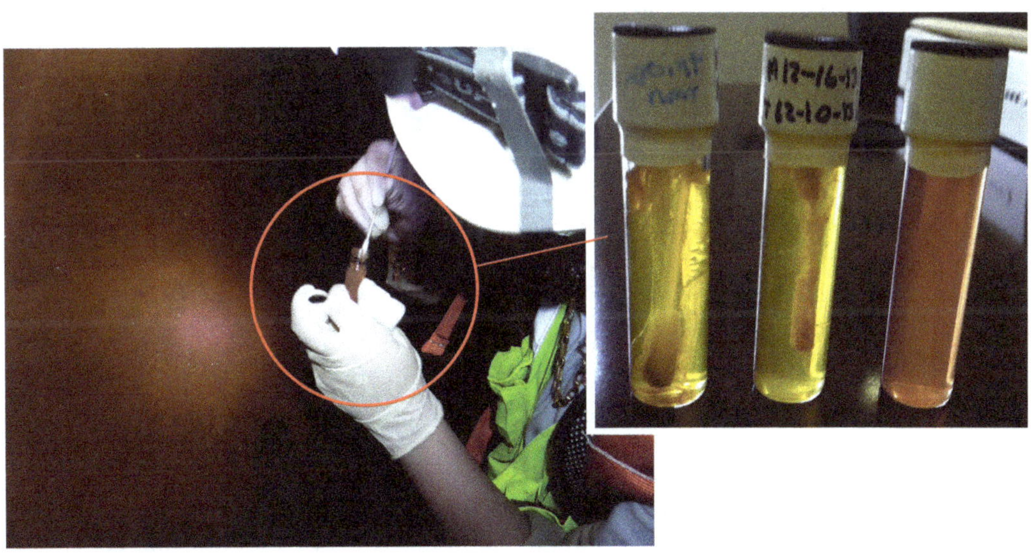

Source: Mersedeh Akhoondan, HDR
MIC: microbiologically influenced corrosion

Figure 6-6 Sampling slime for MIC testing

Laboratory Testing

Chemical analysis of pipe materials, coating, or surrounding soil or water. It is common to collect samples of pipe materials, coating, soil, or water for chemical analysis as part of spot assessments. Samples are typically tested for the presence of soluble salts (California Test Method [CTM] 422 [2014], CTM 417 [2014], ASTM D4327 [2017], ASTM D6919 [2017], American Association of State Highway and Transportation Officials [AASHTO, 1997] T 260), pH, as-received and saturated resistivities (ASTM G187 [2018]), sulfide content (ANSI/AWWA C105 [2010] Appendix A, EPA 376.1), oxidation reduction potential (ASTM G200 [2014]), and many other constituents and properties depending on the type of materials and objectives of testing.

While the details and objectives of each test are beyond the scope of this manual, the three primary chemical tests conducted are for water-soluble chloride and sulfate contents and pH. These tests apply to concrete, cement mortar, soil, and water. Chloride ions are particularly corrosive to ferrous metals, and in higher concentrations, chloride ions can overcome the corrosion-inhibiting effect of concrete on reinforcing steel and other embedded metals. High concentrations of sulfate ions can react with components in concrete to cause concrete degradation and reduced strength in a mechanism known as sulfate attack. Low pH (acid) conditions are deleterious to metals and concrete. High pH (basic) conditions are generally beneficial for iron and steel, but caustics can be damaging to lead, copper, aluminum, zinc, and chromium materials.

It is always recommended to have clean, sealable plastic bags available for collecting samples during any spot assessment of a pipe. While laboratory testing may not be initially planned, the sampling costs little and provides opportunities for future testing. Items to be collected include soil (at various locations within the trench), chipped mortar or concrete, and groundwater. Sampling should also be considered during repairs to the main and at other times when an opportunity exists to get near the pipe.

Elemental analysis. Energy-dispersive spectrometry (EDS) can be used to determine the elemental composition of any region on a material specimen. In this method, EDS detectors measure the pulse height (intensity peak) recorded for an incident proton; such peaks are proportional to the energy of photon responsible for the pulse, as each individual element emits a certain amount of energy. The solid-state X-ray detectors that are used for EDS analysis are typically installed on a scanning electron microscope (SEM) instrument. Therefore, it is typical to refer to this analysis as SEM/EDS testing.

EDS is typically performed on pipe coupons to determine the following:

- the depth of degradation and corrosion at the cross section of the pipe coupon by simply observing the oxygen (the presence of oxygen is typically an indication of corrosion);
- the corrosion products, where the cause and mechanism of corrosion are being investigated as part of the project;
- the composition of pipe or coating materials if there are ambiguities;
- the depth of degradation where certain substances leach out of the pipe matrix, reducing the pipe strength (A common example is the lime leaching process of asbestos cement pipe where calcium leaches out of the concrete matrix, leaving the pipe weakened. When EDS is performed at the cross section of the pipe coupon, the low count of calcium ions near the surfaces suggests that calcium leaching has occurred. This is helpful in determining the depth of degradation and remaining service life of the pipe.); and
- the presence of harmful substances, such as lead.

In EDS testing, generally only a few elements of interest are investigated. In the example shown in Figure 6-7, seven elements are analyzed, and results are reported as relative percent weights. Although these percent weights add to 100 percent, this is misleading. Other elements were likely present, but not measured or reported. Thus, when comparing results from different EDS tests, it is important to recognize differences in the elements that were measured.

Microscopic analysis (optical, nonoptical). Microscopic analysis of samples (i.e., pipe or coating materials) with high-resolution optical microscopes (some with three-dimensional depth-profiling capabilities) is used to detect small defects and abnormalities not visible to the naked eye (Figure 6-8). Greater resolutions and magnifications can be achieved using SEM. SEM is capable of taking high-resolution microscopic photos of surfaces as a result of the interaction between a focused beam of high energy (electron beam) with a solid sample (surface atoms).

Transmission electron microscopy (TEM) also uses a high-energy electron beam. It is more time-intensive than many other frequently used analytical tools. A variety of signals is accessible from these techniques, making it possible to perform chemical analysis at the nano scale. In addition to generating high-image resolutions, it is possible to characterize the crystallographic phase and crystallographic orientation (using electron diffraction experiments), generate elemental maps, and acquire images highlighting elemental contrast.

X-ray diffraction (XRD) residual stress analysis is a method used to investigate failures associated with overloading, stress corrosion, fatigue, stress concentrations, or inappropriate manufacture.

Observing structural features as small as a single column of atoms, determining crystalline structure of nano/micro-scale objects, and determining chemical compositions and chemical bonding structure are sophisticated means of performing failure analysis and identifying defects in the materials.

EDS

Elt.	Line	Error 2-sig	Atomic %	Conc	Units	Calib File	K-Ratio	Z	A	F
C	Ka	0.747	40.485	24.532	wt.%		0.0791	1.0788	0.1471	1.0000
O	Ka	1.173	40.635	32.800	wt.%		0.1879	1.0574	0.2666	1.0003
Si	Ka	0.812	1.061	1.504	wt.%		0.0204	0.9932	0.6649	1.0101
S	Ka	1.660	5.808	9.396	wt.%		0.1585	0.9821	0.8379	1.0089
Cl	Ka	1.095	1.936	3.462	wt.%		0.0549	0.9386	0.8265	1.0071
Ca	Ka	0.497	0.098	0.198	wt.%		0.0038	0.9606	0.9445	1.0279
Fe	Ka	1.894	9.977	28.109	wt.%		0.4955	0.8623	1.0062	1.0000
			100.000	100.000	wt.%					

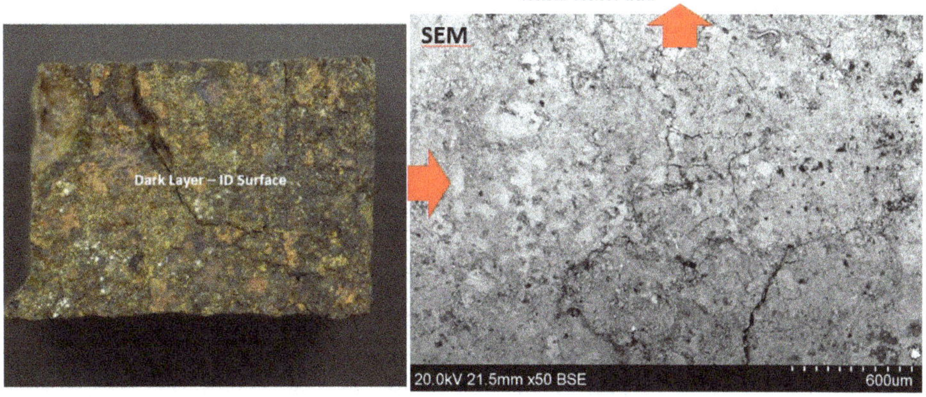

Source: Mersedeh Akhoondan, HDR
EDS: energy-dispersive spectrometry, SEM: scanning electron microscope

Figure 6-7 Identifying corrosion products using SEM/EDS

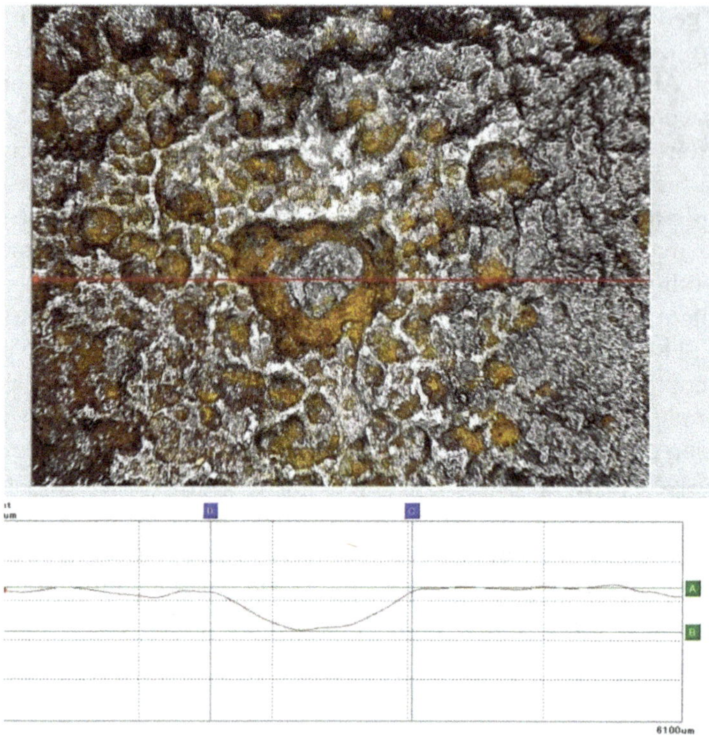

Source: HDR

Figure 6-8 Optical microscopic pitted surface (top); pit depth profiling (bottom)

Metallographic analysis. The structures of metals (i.e., the orientation of grain boundaries, presence of inclusions, and phase compositions) significantly affect their corrosion-resistance properties. For example, sulfide inclusions have a strong role in initiation of pitting in stainless steel. Likewise, grain boundary segregations of carbon induce intergranular stresses and corrosion cracking of carbon steel.

Metallographic evaluations of weld regions to determine the quality of a weld and the risk of corrosion due to sensitization of heat-treated zones are commonly performed (Figure 6-9). Metallographic analyses are highly recommended on pipe coupons when the cause and mechanism of failure are being investigated. Metallographic analysis is relatively inexpensive compared to the cost of excavation.

Here is an example of where metallographic analysis was used effectively. The brittle failure of ductile-iron pipe manifested as longitudinal cracks. While this is common for gray cast iron, it is unexpected for ductile iron. When such brittle failures occurred on one utility's ductile-iron pipeline, a coupon was extracted as part of spot assessment and sent out for failure analysis, including metallographic analysis. The analysis determined that the full spheroidization of graphite had not occurred during the manufacture of this ductile- iron pipe (reducing the material's ductility). Partial inoculation and partial spheroidization of graphite in ductile iron was a common manufacturing defect that occurred during the production of early ductile-iron pipe.

Petrographic examination. Petrographic examination (per ASTM C856 [2018]) of cement-based materials (i.e., concrete, mortar, and asbestos cement pipe) is similar to metallographic analysis of metals where the structure of the matrix can be closely studied for the presence of any abnormalities. This analysis is commonly used to determine if sulfate attack has occurred to the concrete and the degree of carbonation that has occurred.

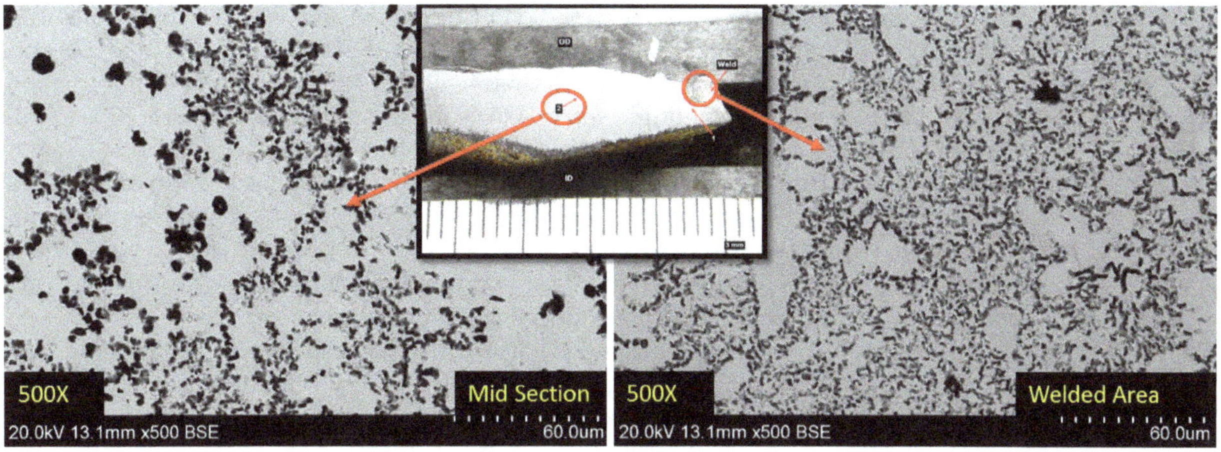

Source: HDR

Figure 6-9 Comparative metallographic analysis of heat-treated (welded) zones

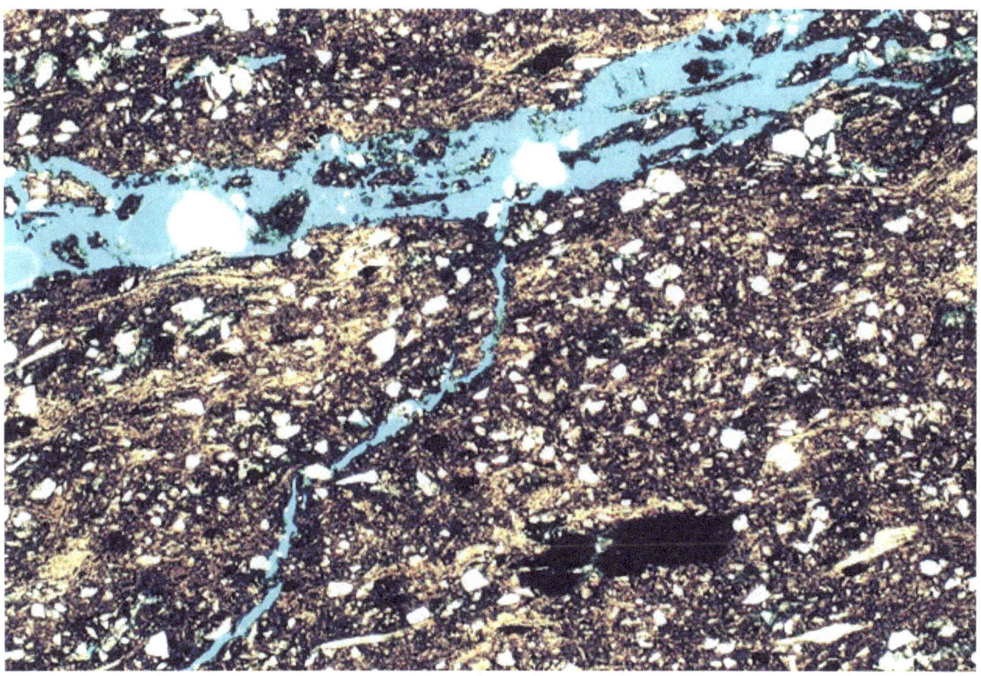

Source: HDR

Figure 6-10 Petrographic analysis of concrete, illustrating cracks

Carbonation is a reaction that occurs between free lime ($Ca(OH)_2$) in the concrete and carbonic acid created by atmosphere carbon dioxide. When carbonation occurs, the pH of the concrete changes from approximately 12 to a more neutral value, negating much of the corrosion protection it provides to embedded or underlying iron or steel. Figure 6-10 depicts petrographic analysis of concrete.

Petrography can also approximate the original water-to-cement ratio of the concrete. This parameter is necessary to correctly model the Fickian transport of chloride ions through concrete. Using this model, the time to initiation of corrosion of reinforcing steel can be determined.

Simulative analysis. Data for corrosion of ferrous metals in a variety of soils were compiled by Melvin Romanoff of the National Bureau of Standards in Circular 579 entitled *Underground Corrosion* (1957). This database, along with other available information, may be used to conduct a similitude analysis to estimate the corrosion rates of pipe in soil. The basic methodology is to identify the representative soil characteristics most likely to be encountered at the site and then use the data presented in Circular 579 to calculate the corrosion rates based on the similitude between the soils documented and the soils anticipated at the site. This analysis assumes that pipes are bare and are installed in undisturbed soil. It also assumes that corrosion occurs at a constant rate; in reality, corrosion generally slows as the products of corrosion help to shield the metal from corrodents.

Accelerated corrosion testing. Accelerated corrosion tests provide a short-term, controlled corrosive condition to obtain information regarding long-term durability performance and corrosion-resistance properties of metals or coating products. One of the best-known accelerated corrosion tests is the salt spray test per ASTM B117 (2018; see Figure 6-11). This method is commonly specified by military, aircraft, aerospace, automotive, and other industries to estimate component service life.

Polarization tests (such as a potentiodynamic test per ASTM G61 [2018]) require special equipment to apply a continuous changing voltage and a current readout. Such tests help to determine the susceptibility of metals to initiation of localized corrosion. While this test is not typically performed as part of spot assessment, it merits consideration, particularly in the situation where service life estimation of pipe, coating, and fastener materials is crucial and relatively few field coupons are available for testing.

Phenolphthalein stain testing. Field tests using phenolphthalein solution were described earlier. This method is also helpful for roughly determining the extent of degradation of asbestos cement pipe, mortar linings, mortar coatings, and other concrete products in the laboratory.

The test reveals where the high-pH of the concrete has been neutralized due to a reaction with atmospheric carbon dioxide and/or the leaching of calcium from the concrete. To perform these tests, it is important to make a fresh cut or to grind the surface of the concrete; otherwise, no reaction may occur. For asbestos cement pipe, such preparation must be performed in a way that does not result in exposure of persons to fibers or contamination of the laboratory.

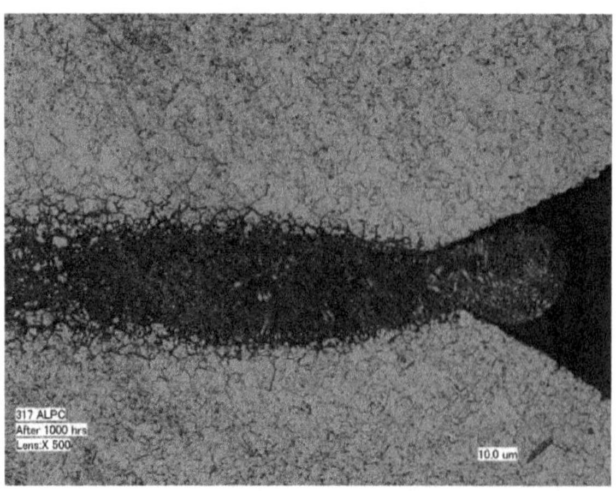

Source: HDR

Figure 6-11 ASTM B117 test on heat-treated areas of stainless steel (1000 h of testing in salt chamber)

Adapted with permission from HDR

Figure 6-12 Phenolphthalein stain testing of asbestos cement pipe

An example of stain testing performed on asbestos cement pipe is shown in Figure 6-12. In this case, over half the cross section of the pipe is "carbonated," meaning that the high pH of the asbestos cement concrete has been neutralized. However, carbonation, in the absence of calcium leaching, does not necessarily weaken the pipe, so interpretation of results may require EDS or other testing. In nearly all cases where carbonation occurs on the inside of the pipe, calcium leaching also occurs and the pipe is weakened, but carbonation on the pipe exterior often is not accompanied by significant calcium loss or weakening of the pipe. This subject is discussed in Water Research Foundation Report 4480, "Development of an Effective Strategy for Asbestos Cement Pipe" (2015).

For mortar-lined or mortar-coated steel and iron pipes or for concrete pipes reinforced with steel or prestressing wires, the depth of carbonation is a concern, even if calcium loss has not occurred. The high-pH environment created by concrete and mortar are main reasons these materials are used for corrosion protection. When carbonation has neutralized the concrete, corrosion of the steel can begin.

POST ASSESSMENT AND INTERPRETATION OF RESULTS

Data are only useful when properly interpreted. When various tests are being performed as part of an evaluation, proper comparison and correlations are needed to draw conclusions. In an ideal world, all test results and collected data are perfectly consistent and agree with the literature, but this is rarely the case. Determining which data and results should be trusted and used without being biased is not always easy. Moreover, determining whether the data collected from one spot are representative of the entire system is another challenge in post assessment and result interpretation.

To evaluate the quality of data in terms of precision, accuracy, representativeness, comparability, completeness, and sensitivity in relation to the intended use, all collected data should first be carefully evaluated, patterns and correlations should be found, and random variations and outliers should be identified. Any conclusions drawn must be supported by the overall data.

Feasibility versus Data Resolution

Because of the unpredictable nature of corrosion, the limitations of analytical methods, and the inherent errors associated with any sampling and analysis procedure, the results of any condition assessment technique may contain an element of uncertainty and therefore may not be representative of the condition of the entire system. An evaluation of the quality of the analytical data in relation to the intended use is crucial in order to make decisions that are supported by data of known and sufficient quality.

To reduce uncertainties and improve data quality, a minimum of three independent tests should be performed as part of spot assessment. Visual inspection can be considered one of the test methods. Others are as described previously. If the overall results of all three test methods do not agree, then additional testing may be needed.

For example, during the spot assessment of a reinforced concrete pipe, no rust staining or other signs of reinforcing steel corrosion were observed. However, potential mapping on the surface of the concrete indicated areas of active corrosion. Samples of the surrounding soil also indicated high concentrations of chloride ions. If conclusions were drawn based only on the appearance of the pipe (visual inspection), one would have concluded that the pipe was in excellent condition. Yet the reality was that localized corrosion of steel beneath the concrete had probably started but had not progressed enough to become apparent. If a fourth test, such as aural testing (sound testing of concrete), were performed, the concrete might sound hollow at those areas of corrosion. Additional tests, such as in situ corrosion rate testing (i.e., EIS or LPR), could also be performed. These would help estimate the corrosion rates and provide insight into how fast corrosion would appear and when failure might be expected. Ideally, a small concrete sample would be removed for petrographic examination or to run elemental and microscopic analysis to verify the presence of corrosion products.

This example illustrates that additional testing can nearly always be conducted. However, a balance is required between the desired quality of information, the cost of testing, and its feasibility. Had the investigators been satisfied only with visual inspections, their conclusions would have been inaccurate. At the same time, the need to perform more elaborate tests such as elemental analysis of the reinforcing steel and petrographic analysis of the concrete may be unnecessary.

Representability of Findings

In many studies, a limited number of samples is used to represent a large population. In terms of applied statistics, a population includes all the elements from a data set (e.g., many feet of pipe) and a sample consists of one or more observations (spot assessments) from the population. A measurable characteristic of a population, such as a mean or standard deviation, is called a parameter, and a measurable characteristic of a sample is called a statistic. While increasing the number of samples (the number of spot assessments) increases confidence in the findings, it does not guarantee that they accurately represent the population.

It is important, therefore, to recognize the limitations of spot assessments, where only a few feet of pipe are examined along a pipeline that spans thousands of feet. When exposing a pipeline for spot assessment, the excavations can be expensive and difficult, so generally only a few samples are practical. Rather than place these excavations at random locations, it is usually best to locate them where corrosion is most likely to be found. However, if sampling is done at these worst-case locations, such samples are—by definition—not representative.

Given these limitations, the findings from spot condition assessment should be presented as statistics and not population parameters. Caution should be used in extrapolating results from a few observations to the entire pipeline.

REFERENCES

AASHTO (American Association of State Highway and Transportation Officials). 1997. T 260—Standard Method of Test for Sampling and Testing for Chloride Ion in Concrete and Concrete Raw Materials. Washington, D.C.: AASHTO.

ANSI/AWWA (American National Standards Institute/American Water Works Association). 2010. C105. Polyethylene Encasement for Ductile-Iron Pipe Systems. Denver: AWWA.

ASME (American Society of Mechanical Engineers). 2012. B31G—*Manual for Determining the Remaining Strength of Corroded Pipelines*. New York: ASME.

ASTM International. 2018. B117. Standard Practice for Operating Salt Spray (Fog) Apparatus. West Conshohocken, Pa.: ASTM International.

ASTM International. 2018. C805. Standard Test Method for Rebound Number of Hardened Concrete. West Conshohocken, Pa.: ASTM International.

ASTM International. 2018. C856. Standard Practice for Petrographic Examination of Hardened Concrete. West Conshohocken, Pa.: ASTM International.

ASTM International. 2018. D6677. Standard Test Method for Evaluating Adhesion by Knife. West Conshohocken, Pa.: ASTM International.

ASTM International. 2018. G61. Standard Test Method for Conducting Cyclic Potentiodynamic Polarization Measurements for Localized Corrosion Susceptibility of Iron-, Nickel-, or Cobalt-Based Alloys. West Conshohocken, Pa.: ASTM International.

ASTM International. 2018. G187. Standard Test Method for Measurement of Soil Resistivity Using the Two-Electrode Soil Box Method. West Conshohocken, Pa.: ASTM International.

ASTM International. 2017. D3359. Standard Test Methods for Rating Adhesion by Tape Test. West Conshohocken, Pa.: ASTM International.

ASTM International. 2017. D4138. Standard Practices for Measurement of Dry Film Thickness of Protective Coating Systems by Destructive, Cross-Sectioning Means. West Conshohocken, Pa.: ASTM International.

ASTM International. 2017. D4327. Standard Test Method for Anions in Water by Suppressed Ion Chromatography. West Conshohocken, Pa.: ASTM International.

ASTM International. 2017. D4541. Standard Test Method for Pull-Off Strength of Coatings Using Portable Adhesion Testers. West Conshohocken, Pa.: ASTM International.

ASTM International. 2017. D6919. Standard Test Method for Determination of Dissolved Alkali and Alkaline Earth Cations and Ammonium in Water and Wastewater by Ion Chromatography. West Conshohocken, Pa.: ASTM International.

ASTM International. 2015. C876. Standard Test Method for Corrosion Potentials of Uncoated Reinforcing Steel in Concrete. West Conshohocken, Pa.: ASTM International.

ASTM International. 2015. D5162. Standard Practice for Discontinuity (Holiday) Testing of Nonconductive Protective Coating on Metallic Substrates. West Conshohocken, Pa.: ASTM International.

ASTM International. 2015. G102. Standard Practice for Calculation of Corrosion Rates and Related Information from Electrochemical Measurements. West Conshohocken, Pa.: ASTM International.

ASTM International. 2014. G3. Standard Practice for Conventions Applicable to Electrochemical Measurements in Corrosion Testing. West Conshohocken, Pa.: ASTM International.

ASTM International. 2014. G5. Standard Reference Test Method for Making Potentiodynamic Anodic Polarization Measurements. West Conshohocken, Pa.: ASTM International.

ASTM International. 2014. G59. Standard Test Method for Conducting Potentiodynamic Polarization Resistance Measurements. West Conshohocken, Pa.: ASTM International.

ASTM International. 2014. G200. Standard Test Method for Measurement of Oxidation-Reduction Potential (ORP) of Soil. West Conshohocken, Pa.: ASTM International.

ASTM International. 2013. D7091. Standard Practice for Nondestructive Measurement of Dry Film Thickness of Nonmagnetic Coatings Applied to Ferrous Metals and Nonmagnetic, Nonconductive Coatings Applied to Non-Ferrous Metals. West Conshohocken, Pa.: ASTM International.

ASTM International. 2012. C1152. Standard Test Method for Acid-Soluble Chloride in Mortar and Concrete. West Conshohocken, Pa.: ASTM International.

ASTM International. 1993. STP1188. Electrochemical Impedance: Analysis and Interpretation. West Conshohocken, Pa.: ASTM International.

CA DOT (California Department of Transportation). 2014. CTM (California Test Method) 417. Method of Testing Soils and Waters for Sulfate Content. Sacramento, Calif.: CA DOT.

CA DOT. 2014. CTM 422. Method of Testing Soils and Waters for Chloride Content. Sacramento, Calif.: CA DOT.

Romanoff, M. 1957. Underground Corrosion. National Bureau of Standards (NBS) Circular 579. Washington: US Department of Commerce.

WRF (Water Research Foundation) 2015. 4480. Development of an Effective Strategy for Asbestos Cement Pipe. Denver: WRF.

WRF. 2012. 4233. Best Practices Manual for Prestressed Concrete Cylinder Pipe Condition Assessment. Denver: WRF.

Chapter 7

Leak Detection

Ahmad Habibian, Chapter Lead, CDM Smith
Andi Corrao, infrastructureMD, Inc.
Dave Johnston, Echologics, Division of Mueller
Emma Quail, Pure Technologies
Paul Meschino, Utility Service Group
Mark Grabowski, Electro Scan
Elly Perets, Utilis
George Kunkel, Kunkel Water Efficiency Consulting

This chapter introduces leak detection techniques that are available in the marketplace and highlights their range of applications, advantages, and limitations. The leak detection technologies covered include internal and external acoustic techniques and various non-acoustic techniques, including ground penetrating radar, infrared thermography, electrical scanning, gas tracer method, and satellite monitoring. Because AWWA M36, *Water Audits and Loss Control Programs*, covers some of these techniques in detail, this chapter limits the coverage of those techniques and expands on other techniques which are not covered in AWWA M36. The chapter also provides guidance on leak detection implementation strategies.

Leak detection is not only beneficial in identifying sources of water loss but can also be used as a performance indicator of the piping system. High leakage can be indicative of corroded pipe, deteriorated joints, and stressed pipe. Unrepaired leaks can further deteriorate the leaking pipe by promoting external corrosion or by washing out pipe bedding, leading to bending and breakage of the pipe. Small leaks often become large leaks or breaks. Research has shown that pipe ruptures are preceded by pipe leaks. It is far better to deal with these problems when the cost and schedule for the repair are manageable. Moreover, through the process of fixing a leak, the pipe is exposed, allowing for external examination and spot assessments (Chapter 6).

Recent advances in acoustic technologies have provided the capability of obtaining information about the structural condition of the pipe wall in addition to finding leaks. Thus, leak detection and a screening-level condition assessment can be conducted simultaneously, as described in Chapter 10.

The costs and system disruptions associated with most leak detection technologies are relatively low. It is prudent to include them in most condition assessment programs. The information provided by performing leak detection, in combination with information provided by more comprehensive assessment technologies described in other chapters of this manual, will provide a more complete picture of the condition of the pipe and the system and will lead to better decisions regarding repair, rehabilitation, or replacement of mains.

WHY LEAK DATA ARE IMPORTANT FOR CONDITION ASSESSMENT

Spot Assessments

After a leak is discovered, a repair crew is dispatched, an excavation is performed, and the leaking pipe is exposed. This presents an opportunity to perform an assessment of the pipe. Whenever a pipe is exposed, spot assessment methods (as described in Chapter 6) should be considered since the cost of the excavation is no longer a factor. Among the data to collect are pipe diameter, pipe material, pipe wall thickness, coating information, type of soil, and the presence and level of groundwater. Soil samples and material samples may also be collected for laboratory analysis. Photos should always be taken.

Failure Analysis

It is also important to understand what caused the pipe to leak and fail. Was the leak on the barrel of the pipe, at a joint, or at a service connection? Was the leak caused by corrosion or third-party damage, or is the cause unknown? Did the pipe break vertically, split horizontally, or experience a rust hole or blowout? Each of these questions leads to different strategies for infrastructure management. There are many guidelines regarding what data need to go into a leak repair report. See Water Research Foundation Report 4490, "Visual Guidance for Common Pipe Failures" (Grigg 2017).

Generally, the collection of failure and condition assessment data is the job of the repair crew, but to collect accurate data, training is important. The crew making the repair is usually focused on completing repairs, often under adverse conditions, and restoring customers back to normal service as soon as possible and therefore may not always gather important data. For catastrophic breaks, or where unusual failures occur, evidence for a forensic analysis should be collected under the guidance of a specially trained technician or engineer.

Using Leaks to Prioritize Pipes for Inspection or Remedial Action

The locations and frequencies of leaks are key criteria for prioritizing mains for more detailed inspection and/or rehabilitation or replacement. As described in Chapter 3, historical repairs are the best predictor of future leaks and breaks. In conjunction with soils data and other factors, these data help a utility assess the risks associated with large and small mains and determine alternatives for management.

Prevention of Larger Breaks

Several Water Research Foundation reports indicate that pipe fractures are often preceded by relatively minor leaks. Rust on the fracture surface and laboratory tests both show that fractures of cast iron and ductile iron often occur after detectable leaks start. While steel pipes are less prone to ruptures, they too will often leak before a more significant failure. Even polyvinyl chloride (PVC) is believed to leak moderately before fracturing. It is therefore prudent to find the minor leaks before they become major problems. This can significantly reduce the cost of the repair but also avoids greater societal and system impacts, as the repairs can be scheduled when convenient.

HOW AND WHEN TO PERFORM LEAK DETECTION

It is recommended that all water utilities provide regular leakage management that includes some form of leak detection at a frequency that is appropriate to their distribution system needs. If the utility has the resources to hire staff, purchase equipment, and issue contracts for training services, it may wish to develop an in-house program in an ongoing or seasonal mode. Generally, this makes sense if the utility is large and leaks occur at a high frequency. Utilities may also consider pooling leak detection resources with another utility.

For most other utilities, however, the services of a leak detection company may be needed. Many leaks are difficult to locate, and leak detection requires both practice and effective, up-to-date equipment. For small water utilities with limited budgets and staff, it is often difficult to maintain the expertise needed. Detecting leaks on large transmission or plastic mains can be particularly difficult, generally requiring the services of a specialty service company.

How often leak detection is performed depends of various factors, including the leak frequency, the cost to produce or import water, the cost of break repairs, and the costs associated with the leak detection work. Utilities can determine leak frequency by applying the technique of Leakage Component Analysis (LCA), a free software tool created by the Water Research Foundation (WRF) as part of the project "Real Loss Component Analysis: A Tool for Economic Water Loss Control" (Sturm et al. 2014). The project report and LCA software tool can be downloaded at www.waterrf.org. This tool complements AWWA's M36 and Free Water Audit Software.

Leak detection is most commonly performed using acoustic methods. These can be deployed using either external devices or in-pipe tools. Non-acoustic techniques are also available. Figure 7-1 lists the techniques covered in this chapter. Because AWWA M36 covers some of these techniques in greater detail, this chapter limits the discussions of those techniques and expands on other techniques that are not covered in AWWA M36. Both manuals should be consulted.

ACOUSTIC LEAK DETECTION

In most water distribution systems, a considerable portion of the leakage occurs as hidden or unreported leaks that produce noise that can be detected aurally (by ear) or by various vibration instruments. How readily leaks are detected depends on the size of the leak, the size of the pipe, the pipe material, the distance between the leak and the detection device, and the level of background noise. Acoustic vibrations do not carry as readily on large pipes or plastic or concrete pipes as they do on small, metallic mains. Other factors that affect detection and accuracy are depth of cover, type of pipe zone backfill, and the presence of paving. Filtering software is helpful in separating the sounds of a leak from normal system and background noise.

This section describes common acoustic leak detection methods—how each method works, its accuracy, and the applicability to distribution pipes and transmission pipes. Acoustic leak detection and different types of leak surveys are also described in detail in Chapter 7 of AWWA M36 (AWWA 2016). M36 focuses on leak detection as a form of water loss control. The primary benefits of an active leak detection program come in the form of

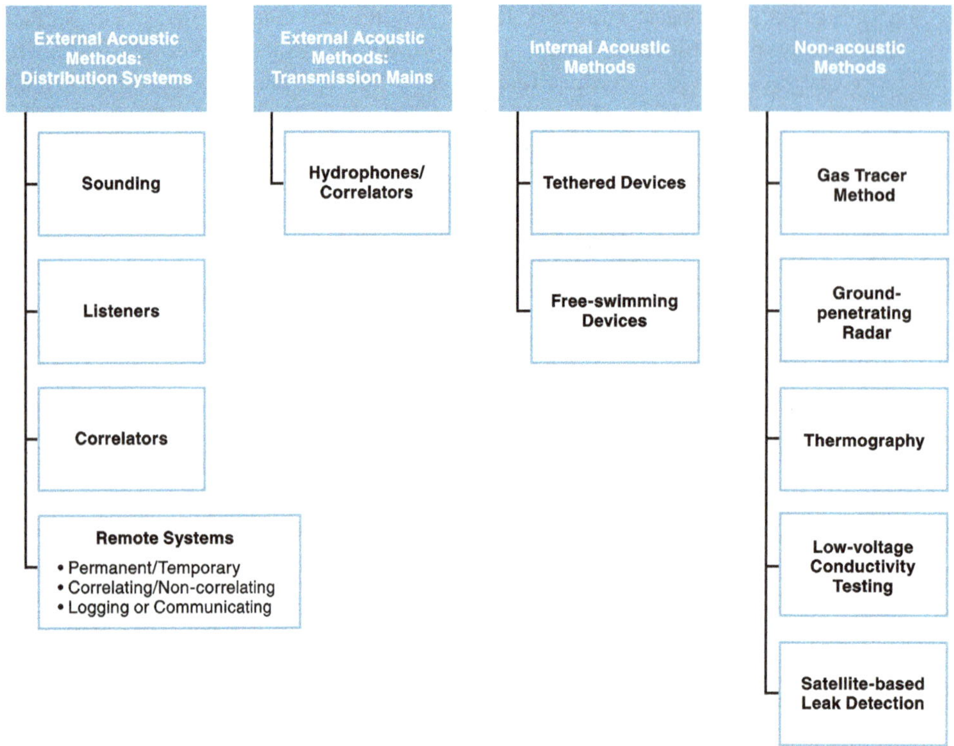

Figure 7-1 Leak detection techniques covered in this chapter

nonrevenue water reduction. However, there are other condition assessment benefits as well, as described previously.

External Acoustic Methods

Various external acoustic-based leak detection methods are available, as listed in Figure 7-1.

Sounding. Acoustic leak detection does not always require expensive electronic equipment. Leaks make noise, and often that noise can be heard using the human ear without the need to amplify or filter the signal. There are tools available to make this easier for field technicians. These tools can be used by a field technician to perform a "sounding survey" on various fittings and appurtenances. The technician listens to the fitting and can gauge the loudness of the signal aurally. The loudness can be graded between 0 and 10, and it can be recorded on a map or into a handheld GPS device.

The basic design of sounding sticks and analog geophones has not changed significantly in many years; however, newer designs with more sophisticated methods have found favor and become more affordable. The newest models optimize materials and resonance chambers to help produce the best possible acoustic signal for the user. The parameters for sounding equipment are indicated in Table 7-1.

These types of equipment are suitable for use in surveys to identify potential areas of leakage. However, they are typically used only as a method to localize potential leaks to within several hundred feet; they cannot be used to pinpoint leak locations except on rare occasions, such as hydrant and valve leaks. Sounding is manually intensive and considerably less reliable at locating leaks than modern leak correlation methods. After a location has been isolated through sounding, more sophisticated tools are required to pinpoint leaks within an acceptable distance for remedial action to be taken.

Table 7-1 Parameters for sounding equipment

Pipe Materials	Diameter Range	Distance Range	Locational Accuracy	Limitations
Metallic, asbestos cement, concrete, composite	Up to 16 in.	500 ft (between listening points) Greater ranges with smaller-diameter pipes and denser or more homogeneous pipe materials	500 ft (between listening points)	Minimum pressure: 15 psi Minimum leak size: 1 gpm

Listeners. Amplified listening devices (listeners) operate on the same principal as the sounding stick. The main difference is that an electronic amplifier is used to enhance the signal before it reaches the user's ear. The method of operation is the same: field technicians place the rod (or tip) of the listening device onto fittings and appurtenances. Vibrations are transferred from the fitting into the rod and then into an amplifier. The signal is amplified, filtered, and sent into a set of headphones where the user can listen to the sound to make a judgment based on the acoustic characteristics. The signal can also be displayed graphically for use and review by the user.

Modern-day listening devices allow technicians to cover more ground, detect, and locate smaller leaks and make better decisions on a day-to-day basis. Although far superior to sounding devices, they are still limited by the judgment and hearing ability of a human operator.

Some models of listening devices also come with ground microphones. These are used to survey the surface above the main between access points. Vibrations from leaks travel down the pipe but are also carried through the ground to the surface. Unless the pipe is covered by a very rigid pavement (concrete) or a very soft surface or is extremely deep, a ground microphone can capture the signal, making it detectable by the human ear. This enables users to pinpoint leak locations to determine where to excavate.

For rigid surfaces like concrete that can mask vibration, utilities can drill a series of holes through the surface material and insert the listening rod to pinpoint the leak. For soft surfaces, the same method can be used; some utilities use a small rigid plate to allow a contact surface for the ground microphone. The parameters for listeners are described in Table 7-2.

Correlators. Leak noise correlators consist of a minimum of two time-synchronized acoustic detectors that use a computer processor to automatically perform an analysis of the signals. By comparing sounds received at the two sensor stations, the devices can calculate the location of the leaks in between the sensors with accuracy within approximately 5 ft. The term correlator stems from an algorithm (called a correlation function) that compares the time of flight between each channel. By comparing the times, the signal's position on the water main can be calculated. It is similar to triangulation except that only two measurements are needed since the problem is two-dimensional. The travel time of the

Table 7-2 Parameters for listeners

Pipe Materials	Diameter Range	Distance Range	Locational Accuracy	Limitations
Metallic, asbestos cement, concrete	Up to 24 in.	750 ft (between listening points) Greater ranges with smaller-diameter pipes and denser or more homogeneous pipe materials	750 ft (within listening points), ~2 ft for ground microphones depending on pipe depth, soil, and surface type	Minimum pressure: 15 psi Minimum leak size: 1 gpm

Table 7-3 Comparison of leak detection correlators

Characteristics	Advantages	Limitations
Digital radio communication	Higher signal-to-noise ratio and sensitivity	Limited transmission distance
Analog radio communication	Longer transmission distance	Can suffer from interference
Central processor	Stand-alone device No external processor such as laptop or notebook required	High cost and limited ability to record and manipulate files
Laptop connectivity	Correlation software may be upgraded on a PC Files can be emailed and shared with colleagues for a second opinion	Requires purchase of a laptop
Multiple sensor types and/or multiple correlation channels	Higher sensitivity on a wider variety of pipe diameter and pipe material, including plastic pipe (polyvinylchloride [PVC] and polyethylene) Can do multiple correlations with one setup; useful for busy intersections, for example	Complicated and more expensive

Table 7-4 Parameters for correlators

Pipe Materials	Diameter Range	Distance Range	Locational Accuracy	Limitations
Metallic, asbestos cement, concrete, composite, PVC, PE	Up to 24 in.	300–1,500 ft, depending on pipe material	<3 ft with accurate data	Minimum pressure: 15 psi Minimum leak size: generally 1 gpm

PE: polyethylene, PVC: polyvinylchloride

signal depends on the distance, the materials, and the sizes of the pipes between the leak and the detector, so it is important to know this information before the leak noise correlation is conducted.

Leak noise correlators are available in many different forms; some use analog radio communication, some use digital; some have built-in computer processing at a central location, some use software installed on laptops; and some have multiple and varied types of sensors, which are chosen depending on pipe type, diameter, and fitting availability. Each option has its own strengths and weaknesses, as described in Table 7-3. When buying a correlator, it is important that the purchaser understand the options and determine which is best for the situation.

Correlators detect acoustic frequencies beyond what a human ear can perceive, so they can identify leaks that traditional listeners or sounding equipment cannot find. Leak noise correlation is far less dependent on the opinion of the operator. This makes correlators superior devices for locating smaller leaks on all pipe materials, particularly pipe materials that transmit sound poorly, such as concrete, PVC, and polyethylene (PE), or large-diameter mains.

Correlators can generally locate leaks to within 5 ft on most water mains, allowing for relatively precise excavations for repair. They are often used for pinpointing leaks after a listening-device survey has been completed. They are also used as the primary leak survey tool on mains where listening devices do not work well, including large transmission pipes. They can also be used as survey-level tools on other pipes that are considered to be of high consequence or any pipe where a more methodical approach is needed. The parameters for correlators are described in Table 7-4.

Leak loggers. Leak loggers are essentially smart listening devices. They are battery-powered single-channel devices that monitor noise daily at preset times from a pipe appurtenance. They are widely used by utilities as a part of a leak detection program. Loggers can be mounted permanently at locations in the distribution system or rotated across the system in "lift-and-shift" mode. When rotated, loggers are frequently deployed in large groups (typically 25 or more at a time). They are typically placed on valve operating nuts or similar pipe connections at intervals ranging from 50 to 500 ft.

Suitable installation locations are generally below ground where a pipe appurtenance is accessible. Traditional loggers are generally not placed on fire hydrants due to risk of theft or vandalism. Loggers are often placed in contact with valve nuts, curb stops, or service lines. Most loggers require manual data collection. This is generally known as "drive-by" data collection, where a crew member is dispatched to collect the data on a recording device. The data are then downloaded at the office for analysis.

The results are similar to those provided in leak listening surveys as described previously. Each logger provides a measure of the acoustic energy from a fitting. This can be displayed in the form of a frequency spectrum or as a score on leakage level. The benefit of using this type of system is that data are collected from many loggers at the same time and displayed on a map for easy analysis. Analysts can review the results and determine whether a leak exists within a few city blocks. This allows for a pinpointing team to be deployed and, using a correlator, the team can isolate the leak to within a few feet. The other advantage is that data are readily collected at night, when interfering background noise is minimized, and collected over an extended period, allowing for better detection of small leaks.

Leak loggers provide a sophisticated method for performing a leak listening survey. Instead of having one person listen to one fitting for a limited amount of time, loggers can be deployed in large numbers to cover wide areas of a distribution network. As with traditional listeners, they do not pinpoint leaks. Table 7-5 provides the parameters for leak loggers.

Remote systems. In recent years, remotely monitored leak logging/correlating systems have become commercially available. These types of systems can be installed temporarily, semipermanently, or permanently. The common factor is that, once installed, these systems communicate information via wireless networks back to a central location. A remote leak detection system can provide utilities with near real-time information about their water mains without deploying technicians to the field.

Remote systems work in a similar fashion to leak loggers as described previously; however, they can offer much more sophisticated analysis and pinpointing abilities. Also, remote monitoring systems allow for early detection of leakage, which enables utilities to identify leaks sooner and better prioritize repairs. Early detection and repair of hidden leaks reduces water loss and reduces repair costs, as leaks are less severe and can be repaired during normal work hours.

As with survey level correlation technology described previously, remote technologies vary greatly depending on what diameter range is in question. For that reason, the discussion will be split into two parts: distribution systems (typically pipes less than 16 in. in diameter) and transmission systems (typically pipes 16 in. in diameter or greater).

Table 7-5 Parameters for leak loggers

Pipe Materials	Diameter Range	Distance Range	Locational Accuracy	Limitations
Metallic, asbestos cement, concrete, composite	Up to 16 in.	500 ft	500 ft	Sensitive to pressure, sensor spacing, leak size, pipe material, and background noise

Source: Matchpoint

Figure 7-2 A leak detection monitor installed in a below-ground chamber

Distribution system monitoring: Typical monitoring systems for distribution systems use the latest generation of acoustic sensors and wireless communications hardware to transmit and relay data to a central location.

The main component of a leak detection monitoring system is the leak detection node (or monitor), which comprises an acoustic sensor, a logger board, communications hardware, and a battery. Some models are installed in below-ground chambers similar to that of traditional loggers, as shown in Figure 7-2. One company connects its sensors to the service line at the meter (see Figure 7-3). Other models have nodes contained in hydrant caps, as shown in Figure 7-4.

Maximum spacings of the monitors are 400–1,000 ft, depending on pipe material and the product manufacturer. Data are communicated by either a cellular or radiofrequency (RF) network, as shown in Figure 7-5.

In addition to the single-channel analysis that is provided by leak loggers, remote systems are trending toward the use of two-channel analysis similar to leak detection correlators. By analyzing the noise collected at a pair of nodes, the location of the leak can be calculated using a computer algorithm, as described previously in this chapter.

Using a cloud or web interface, system operators can view reports from a remote monitoring network. Many systems have dashboards that show locations of interest found by the monitors. Operators can analyze individual channels or view correlation results from pairs of nodes. They can review the information and determine whether crews need to be dispatched immediately or if a location should be monitored for a few more days before a decision is made. Most systems allow for a web-hosted graphical user interface to be integrated into existing geospatial information system portals.

Remote monitoring of distribution systems represents a new era in proactive leak detection. After nodes are installed, operators can monitor the system without having trucks and labor deployed to the field. This allows for faster response time, minimizing loss of water and damage to streets, adjacent utilities, and properties. Currently, remote

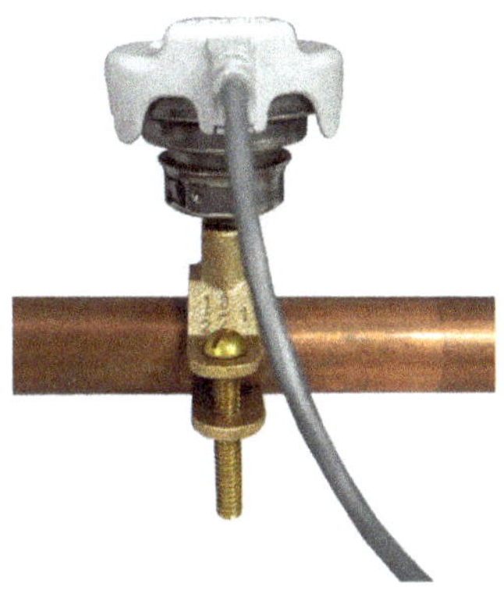

Source: Itron

Figure 7-3 A leak detection sensor connected to a service line at the meter

Photo: Shutterstock.com/3D Vector

Figure 7-4 A leak detection node contained in a hydrant cap

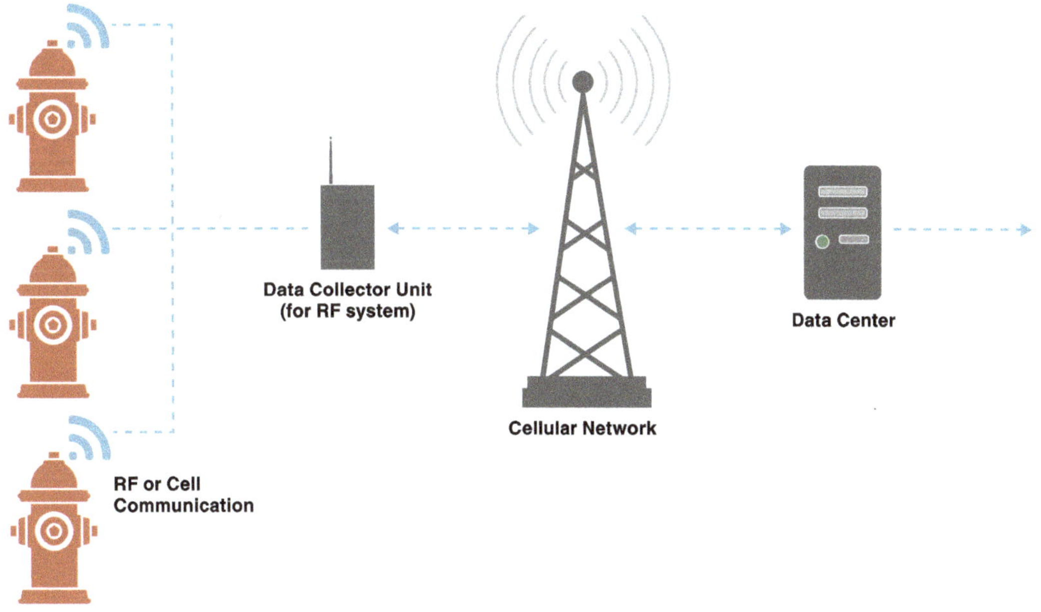

Adapted with permission from Echologics
RF: radiofrequency

Figure 7-5 A typical communications network

leak detection systems using two-channel analysis generally locate leaks within about 15 ft. It is therefore still necessary to use other tools for final pinpointing before the start of excavation.

Permanent remote monitoring systems typically require a notable upfront investment in equipment purchase and system commissioning. The utility must also ensure that a trained analyst reviews the system alerts on a regular basis and initiates appropriate repair actions. The analyst should be trained to recognize and rule out false positive alerts from the system. Finally, the utility should anticipate the life of this smart equipment deployed in a field environment and what the recurring costs may be to replace failed or obsolete listening units every 3–10 years. Parameters for remote correlating leak detection systems for distribution systems are described in Table 7-6.

Transmission main monitoring: Transmission main monitoring systems provide leak detection and additional sensing information for large-diameter transmission mains. Similar to the distribution system monitoring systems, the core component of a typical system for transmission mains is a node. A node consists of a sensor, a data processor, cellular communications hardware, and a power source.

Nodes are typically installed in a chamber where access to the water column can be obtained. Figure 7-6 shows a typical installation in an above-ground pedestal and an overview of a transmission monitoring network. In this example, a hydrophone has been installed on a riser pipe leading to an air release valve. It is most common for hydrophones to be used as the sensor for leak detection. These sensitive devices require connection to the water column, but not necessarily insertion within the main. Where access chambers do not exist at proper spacing, the cost of additional access must be included in the cost of the system.

In addition to leak detection, these systems have additional capacity for measurement of pressure, flow, temperature, and chlorine residual. Most systems are compatible to take other sensing inputs, as long as the output is within a certain range (typically 4–20 mA).

At least two nodes are needed to perform leak detection. The length of pipeline bracketed by a pair of nodes is the portion of the transmission main being monitored. A series of interconnected nodes form the network to monitor the main. Data are automatically captured from the pipeline at scheduled intervals and wirelessly uploaded to a secure server via a cellular network.

The data files are autonomously analyzed, and acoustic correlation is performed. The system identifies potential leaks and other acoustic anomalies in the transmission main and alerts the operator. Following autonomous analysis, data files are uploaded to a server and are further analyzed by the automated analysis software. Data are accessed, and results are displayed on the user interface. Information can be graphically displayed in GIS. The interface is cloud-based, and each user is provided with unique login credentials. Through this interface, the user can

- view the results of analysis,
- be notified of points of interest,
- acknowledge points of interest,
- reprocess data files,
- listen to leak noise, and
- view historic data.

Table 7-6 Parameters for remote correlating leak detection systems: distribution systems

Pipe Materials	Diameter Range	Distance Range	Locational Accuracy	Limitations
Metallic, asbestos cement, concrete, composite	Up to 16 in.	Up to 1,000 ft	<3 ft with accurate data	Pressure, background noise (false positives), equipment life

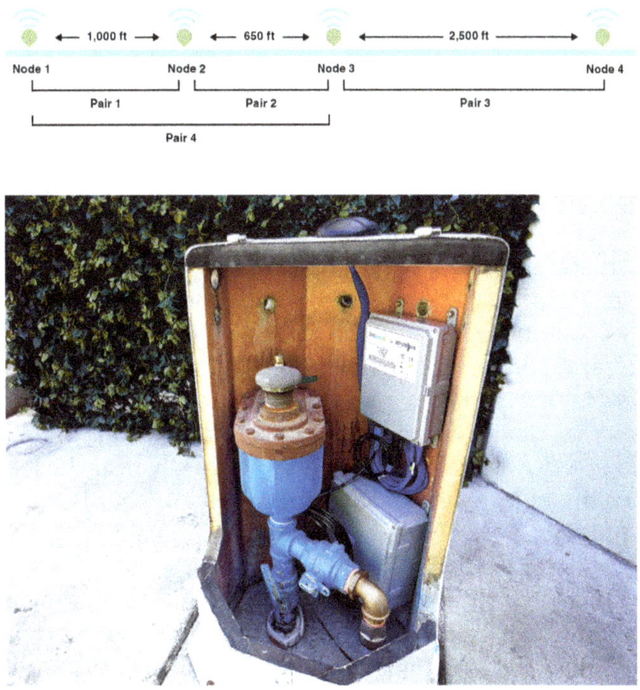

Adapted with permission from Echologics

Figure 7-6 Typical node layout on a transmission main (top) and a node installation in an aboveground pedestal (bottom)

Table 7-7 Parameters for remote leak detection systems: Transmission mains

Pipe Materials	Diameter Range	Distance Range	Locational Accuracy	Limitations
All	Up to 108 in.	2,500 ft	6 ft	Pressure, sensor spacing, appurtenance access

Using acoustic correlation technology to permanently monitor transmission systems is a unique way to monitor pipes for leakage. This method allows for monitoring in remote locations where grid power may not be available. These types of systems provide high locational accuracy on leaks and give asset managers the ability to keep a close eye on their most critical assets. Parameters for remote leak detections systems for transmission mains are described in Table 7-7.

As mentioned earlier, permanent remote monitoring systems typically require a notable upfront investment in equipment purchase and system commissioning, and the nature and extent of such equipment for transmission mains can require a considerable investment. Therefore, this technology may be best designated for the most critical transmission pipelines in the water supply system. The utility must also ensure that a trained analyst reviews the system alerts on a regular basis and initiates appropriate repair actions. The analyst should be trained to recognize and rule out false positive alerts from the system. Finally, the utility should anticipate the life of this smart equipment deployed in a field environment and what the recurring costs may be to replace failed or obsolete listening units every 3–10 years.

Internal Acoustic Methods

Devices used within a water main are often able to get within inches of the leak, which enables detection of smaller leaks, particularly on pipes where sound does not carry for long distances. Detection from within the pipe can also allow for highly precise location of the leak, particularly when the device is held by a tether. Internal leak detection is applicable to both distribution systems and transmission mains.

The sensing device is typically inserted into a pipeline under pressure, using existing pipeline outlets. The device is either propelled by the flow of water or pushed through the pipe to listen for the sound of water leaving the pipe into a lower-pressure environment (i.e., a leak). Traditional external acoustic methods rely on the pipe wall to carry the leak sound and are therefore challenged by a number of variables, as discussed previously. Pipe material, pipe diameter, and the distances between connection points often limit the size of leak that can be detected. In-pipe leak detection technologies, however, are faced with less resistance as the sensor is brought directly near the leak. These devices therefore have the ability to pinpoint smaller leaks with a higher level of confidence than external methods provide. As with acoustic methods in general, operating pressure needs to be 15 psi or higher for the leak noise to be detected.

Operational considerations and inspection specifics dictate which technology will be most suitable to each individual situation. There are two primary types of in-line leak detection technologies available in the market today.

1. Tethered hydrophones
2. Free swimming (untethered) devices

Due to the complexity and expense of the in-line leak detection equipment and the skill level required of the operator, in-line leak detection has nearly always been provided as a contracted service. This may change as devices evolve and increased competition drives down costs.

Tethered leak detection devices. With tethered devices, a sensor containing a hydrophone is inserted into a live pipeline via an existing outlet, such as air valve piping (for transmission mains) or a fire hydrant (for distribution pipes). The device is attached to a cable that is used to control its movement. If a suitable entry point is not available, a tap can be made to accommodate insertion. For long-distance travel in transmission mains, the sensor is propelled by the flow of water using a drag chute that carries the tethered sensor head through the pipeline. The sensor device travel speed is controlled by a copper or reinforced fiber optic cable that also transmits acoustic (and optional closed-circuit television [CCTV]) data back to the exterior data collection unit. In smaller distribution pipes, a shorter but stiffer cable can be used to push the device through the pipe for a distance up to 300 ft. This type of inspection is "real time," and, in addition to the hydrophone used to detect leaks, the sensor also may contain a camera for CCTV inspection. CCTV can show the general condition of the pipe interior, reveal air pockets, and show valves and fittings.

The tethered system offers the operator increased control over the inspection with the ability to position the device over particular points of interest and visually assess the condition of the pipe at the time of inspection. This feature significantly increases the accuracy of the sensor and the sensitivity; therefore, very small leaks—as small as 0.005 gpm—can be located. It also has the ability to identify multiple leaks in one inspection along the suspect section of pipe. Pipe material does not dictate the operability of a tethered system; however, pipe diameter is somewhat restrictive from an operational perspective, as the pipe must be large enough to allow the sensor to flow freely through the pipe. This particular system is adaptable to complex urban networks due to high levels of ambient noise and the level of control over the device. Table 7-8 lists the operating parameters for tethered devices. Figure 7-7 shows a typical tethered leak detection insertion setup for a long-distance inspection.

Free-swimming devices. Free-swimming leak detection devices are inserted into a pressurized pipeline via a new or existing valved outlet, such as a sufficiently sized tap or fire hydrant. The device traverses the pipeline at a velocity proportional to the rate of flow. If flow conditions are optimum, the sensor can travel for a significantly long distance, navigating gate valves and butterfly valves without limitation. It is important to be aware of flow direction, as the device will follow the most significant flow. As the sensor moves through the water column, it records acoustic data using a hydrophone. The sensor can be retrieved at another opening downstream with a net-like device. After the sensor is removed, the data are downloaded and analyzed to determine whether leaks are present. The acoustic signature of a leak varies with leak type and size. Figure 7-8 is an illustration of a free-swimming leak detection device from insertion to extraction.

The performance of this type of leak detection technology is not restricted by pipe material or pipe diameter, as the same sensor is adaptable to all situations. However, from an operational standpoint, the pipe has to support the size of the device, as it must be able to move freely along the pipe without restrictions. To track the device's location, receivers are attached to the exterior of the pipe at multiple locations, often at existing appurtenances.

Table 7-8 Parameters for tethered devices

Pipe Materials	Diameter Range	Distance Range	Locational Accuracy	Limitations
All	10 in.+ for long distances; 4 in. and greater for push-cable inspection	4,750 ft (0.9 mi) for long-distance inspection propelled by the flow of water; 300 ft for push-cable inspection Cable can be retracted and advanced multiple times to better discern the leak noise strength and location	±1.5 ft.	Pressure: 15–200 psi Flow velocity: 1–10 ft/s for long-distance inspection Bends (the number and degree of bends will limit travel distance) Minimum 2-in. insertion port Will not navigate butterfly valves or mostly-closed gate valves Best to isolate pipelines branching from the test pipeline

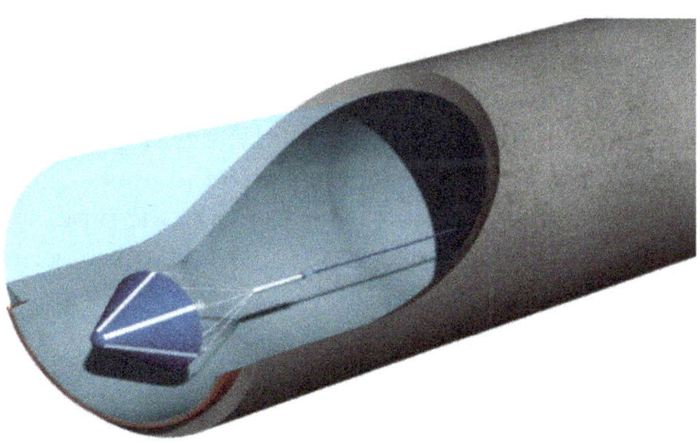

Source: Pure Technologies

Figure 7-7 Tethered leak detection tool

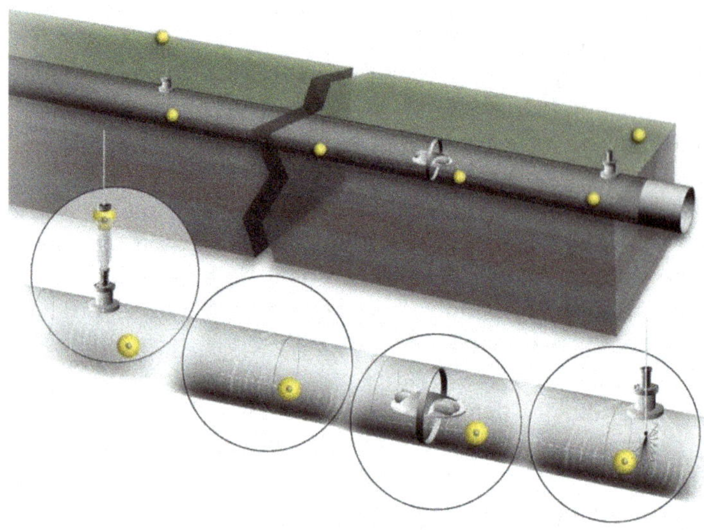

Source: Pure Technologies

Figure 7-8 Free-swimming leak detection device

The section of pipe to be inspected must be isolated where there are "known" laterals, so that the device does not go astray. All main-line valves along the course of the inspection must be fully opened so that there are no obstructions and the device can travel freely. Butterfly valves and gate valves can be easily traversed as long as they are open. The receivers placed on the exterior of the pipe (approximately every 3,000 ft) are used to track the location of the device. They do not record data. The free-swimming device uses an internal hydrophone to listen for the acoustic signature of leak sounds, and the data are stored on board for analysis after inspection.

The data that are captured in the free-swimming device are interpreted by special data analysts who review the data visually and by the acoustic signatures of the recorded noise. These data will indicate distance of the leak from a milestone along the pipeline, and analysis will determine whether the leaks stem from a joint or from within the pipe barrel.

Table 7-9 lists the operating parameters for free-swimming leak detection devices. Increased operational pressure can amplify the leak sound; however, unlike traditional external methods, the sensor is brought directly to the site of the leak, enabling smaller leaks to be detected and accuracy to be increased. With conventional (external) leak detection technologies, the sound of the leak must travel to the sensor outside the pipe; therefore, operational pressure and an array of other challenges affect leak detection accuracy. External sensors must compete with background noise, pipe material, availability of access points, distance between access points, depth of the pipe, soil composition, pipe diameter, and the size of the leak, to name but a few. On the other hand, internal leak detection is not affected by these limitations.

While free-swimming devices are primarily used for leak detection, some devices available in the marketplace can also carry multiple additional sensors such as a CCTV camera for video inspection and a pressure recorder for leak location correlation during data analysis. Devices are capable of reporting general leak locations to pipe "clock positioning" and distance orientation along the pipe where video inspection is being recorded. Buoyantly balanced devices ensure pipe-centric movement with the flow where velocities are strongest and more constant.

Table 7-9 Parameters for free-swimming devices

Pipe Materials	Diameter Range	Distance Range	Locational Accuracy	Limitations
All	6 in. and greater	21 h of recording time; several mi	±6 ft	Pressure: 15–500 psi Flow velocity: 0.5–6 ft/s Minimum 4 in. insertion and extraction ports Once-through sounding recording; cannot repeat sounding to better discern the leak noise Best to isolate pipelines branching from the test pipeline

NON-ACOUSTIC LEAK DETECTION

Acoustic leak detection—from either outside or inside the pipe—is just one means of detecting leaks in pressurized water pipes. Several other techniques have been developed to identify leaks in distribution and transmission system piping where leak noises are weak or difficult to detect. While each of these techniques has its advantages, each technique also has limitations. These techniques are in limited use commercially, but research continues on these and other new methods. Some of the methods included in this section are addressed in further detail in AWWA M36.

Gas Tracer Method

This leak detection method is best applied where there is a known leak that cannot be pinpointed. The principle behind the gas tracer technology is illustrated in Figure 7-9. With this technology, helium gas (usually mixed with nitrogen) is injected into live water mains, with no need to isolate a zone, depressurize the system, or shut the system down. Once the gas is injected into the water main, it mixes with the flowing water and travels throughout the pipe network to the desired area(s) for survey. When the helium-infused water leaves the pipe, the helium separates from the water. Since helium is five times lighter than air, the helium floats to the earth's surface where it can be detected and measured with specialized monitoring equipment. Elevated levels of helium detected above a leaking pipeline indicate a leak is nearby. As a noble gas, helium is absolutely harmless, adding no taste or odor to the water. It cannot react with water, anything in the water, the pipe materials, or any other substance.

The gas tracer method is suited for any pipe material and any diameter. One advantage it has over other leak detection methods is its ability to find and locate leaks in

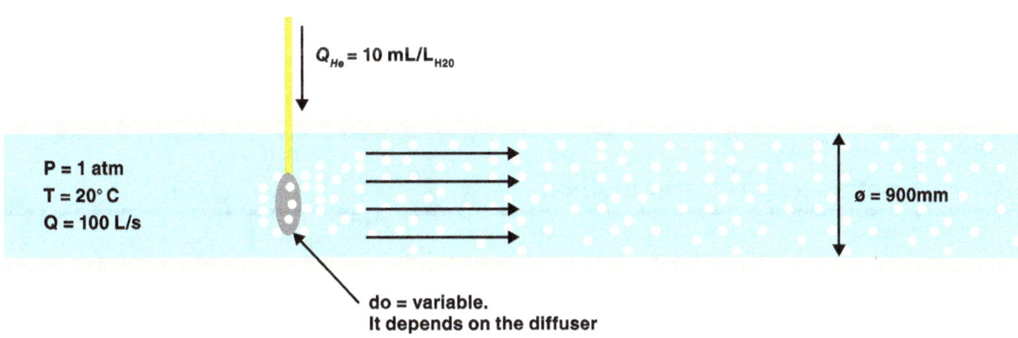

Adapted with permission from Utility Service Group

Figure 7-9 Gas tracer leak detection method

nonmetallic and large-diameter pipes. In nonmetallic pressure pipes (PVC, PE, or concrete), noise does not travel readily and is difficult to detect with external acoustic leak detection technologies. The gas tracer method is effective in large-diameter mains, where there are very few or distant listening points and where low/intermittent pressures, high pressures, or changing pipe geometry make other methods difficult.

The sensor or "sniffer" must be walked above the pipe, so it is necessary that the pipe location be well marked and readily traversed. Also, if the ground surface is impermeable, there may be a need to drill holes through concrete or asphalt to allow the gas to come to the surface.

The helium gas must be certified per NSF/ANSI 60: Drinking Water Treatment Chemicals—Health Effects. The gas is injected through a standard ¾-in. corporation stop or fire hydrant installed on a water main, pipe, or appurtenance upstream of the area to be surveyed for leaks. Since the handling of gases requires specialized skills and equipment, this technology is not typically provided by in-house water utility personnel but instead by specialized service providers who contract with water utilities to conduct the leak detection survey. Table 7-10 summarizes the pertinent parameters for this leak detection method.

Additional information on this method may be obtained from AWWA M36.

Ground-Penetrating Radar

The ground-penetrating radar (GPR) technique is similar in principle to seismic and ultrasound techniques. A transmitting antenna sends a short-duration pulse of high- frequency electromagnetic energy into the ground. The pulse is partially reflected back to the ground surface by buried objects or voids in the ground or by boundaries between soil layers that have different dielectric properties. Reflected radar signals are captured by a receiving antenna. The ground's interior is scanned with radar waves in a manner similar to that of an ultrasound to obtain cross-sectional images.

GPR could, in principle, be used to detect leaks in water pipes by detecting underground voids created by leaking water as it circulates near the pipe or by detecting anomalies in the pipe depth as measured by radar. Soil that is saturated by leaking water slows down radar waves and makes the pipe appear deeper than it should be. Table 7-11 summarizes the parameters for GPR. GPR detects all types of water in the ground, so its use is limited to areas where the pipe is above the water table and soils that can be saturated with water. In practice, GPR is often combined with thermography (discussed in the next paragraph). Additional information on this method may be obtained from AWWA M36.

Thermography

This technique detects thermal infrared radiation and displays it as visible images. In an infrared radiation image, the ground surface above a leak may appear cooler or warmer than the surface farther away from it. This temperature difference may reflect differences in the temperature of the leaking water and the overlying soil; considerable heat may be transferred between leaking water and surface soil. Also, soil close to the water becomes

Table 7-10 Parameters for gas tracer method

Pipe Materials	Diameter Range	Distance Range	Locational Accuracy	Limitations
All	All	Unlimited—function of time and quantity of gas infused into distribution system	Within 3 ft of indicative leak	Pipe needs to be pressurized Less effective if soil surrounding the pipe is clay, swampy, or frozen Gas may travel along vertical profile of pipe

Table 7-11 Parameters for ground-penetrating radar

Pipe Materials	Diameter Range	Distance Range	Locational Accuracy	Limitations
All	All	Not applicable	5 ft	Soil conditions affect accuracy; sophisticated equipment and user skill required; slow, and requires advanced knowledge of pipe location

Table 7-12 Parameters for thermography

Pipe Materials	Diameter Range	Distance Range	Locational Accuracy	Limitations
All	All	Not applicable	5 ft	Requires sophisticated equipment and user skill Affected by ambient temperatures of air and soil, relative humidity, and seasonal effects Not yet available commercially in an affordable, user-friendly package

saturated by leaking water, which may change its thermal characteristics and make it a more effective heat sink relative to dry soil away from the leak.

A thermographic survey of an area uses a high-resolution, commercial infrared camera system. The camera should be focused on the ground surface and should capture images over a period of time. Thermographic images do not distinguish between the temperature irregularities caused by leaking water or other sources of temperature variance such as underground electric lines or hot springs. There are specialty companies that employ this method in combination with GPR.

Thermography leak detection has limited use in the water industry because separating other heat changes from water is a challenge; also, the cost of flyovers is a consideration. Table 7-12 summarizes the parameters for choosing thermography as a viable leak detection method.

Low-Voltage Conductivity

Low-voltage conductivity testing is a technique that uses electric current to identify leaks in nonmetallic pipe. It is a tethered inspection method in which a probe is inserted into a pressurized water distribution pipe through a fire hydrant (wet or dry barrel) or another access point specifically prepared for launching the device.

The test is carried out by applying an electrical potential (typically in the range of 9–11 V with a frequency of 0.5–1.5 kHz) between an electrode in the pipe and an electrode on the surface. The exterior electrode can be anything that allows for electrical connection to the earth—such as a grounding rod, guy wire anchor, sign pole. A simplified electrical circuit for this procedure is shown in Figure 7-10.

As the electrode moves within the pipe, the location and sizes of leaks are found by the amount of electric current that is measured. Where water is leaking from the electrically insulating pipe, electrical current also leaks out. The high electrical resistance of the nonmetallic pipe wall allows only a very small electrical current to flow between the two electrodes unless there is a hole or void in the pipe wall, such as a crack, defective joint, or faulty connection. The greater the electric current flow through the pipe opening, the larger the size of the defect.

Since low-voltage conductivity testing is based on the difference of the high electrical resistivity properties of the pipe walls (i.e., nonconductive) versus the low electrical resistivity of the earth surrounding the pipe, this test method works well on nonconductive

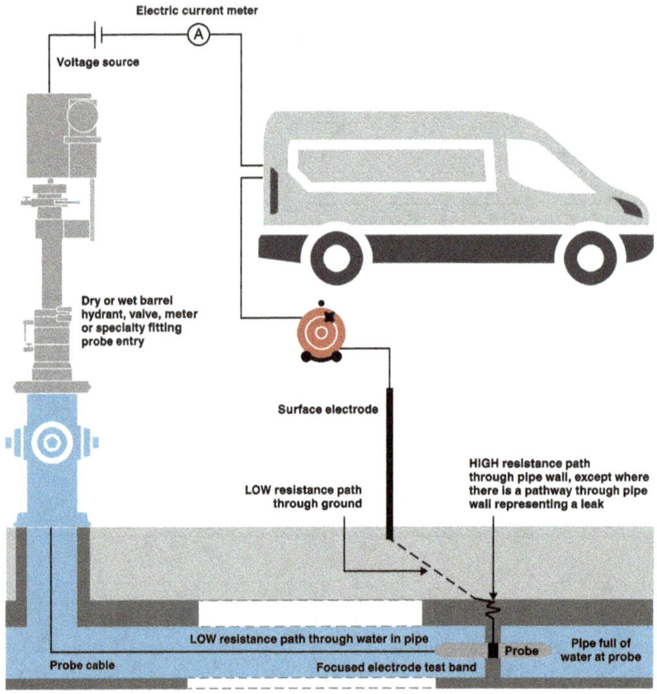

Adapted with permission from Electro Scan

Figure 7-10 Basic diagram of low-voltage conductivity circuitry

pipe materials such as plastic mains (PVC or PE), mains lined with cured-in-place pipe (CIPP), or fiberglass-reinforced pipe (FRP). Externally applied acoustic methods are often less effective on these types of pipe. On the other hand, metallic pipe materials (i.e., ductile iron, cast iron, and steel), which are excellent conductors of electrical current, are not well suited for low-voltage conductivity testing. Use on concrete pipe materials (AC, PCCP, and RCP) is less problematic—accurate results have been demonstrated on concrete pipes. Currently, this technology is only available through the services of a skilled and specially equipped service provider. Table 7-13 summarizes the capabilities and limitations of low-voltage conductivity testing.

Satellite-based Leak Detection

This technique uses earth-orbiting satellites to detect potable water leakage in urban water systems. The technology has been adapted from the search for water on other planets.

Satellite-based leak detection is based on an algorithm that detects underground leaks through the analysis of micro-spectral satellite imagery. The algorithm relies on the fact that different substances in nature have unique spectral signatures. The algorithm processes each radar image pixel and locates the spectral "signature" of freshwater in it. This signature is the relationship between an electromagnetic wavelength and its interaction with water.

Microwave satellite images are received as the input, after which a mathematical manipulation is applied to provide analysis of the leak's coordinates. In addition, the analysis estimates the size of the leak (from as little as 0.03 gpm) and has the ability to detect potable water leaks between 3 and 10 ft underground. At this depth, potable (treated) water does not usually exist without a reason, so after filtering out lakes, swimming pools, drainage assets, sewage assets, and other interferences, this information provides an assessment

Table 7-13 Parameters for low-voltage conductivity testing

Pipe Materials	Diameter Range	Distance Range	Locational Accuracy	Limitations
CIPP, FRP, HDPE, PVC	6–16 in.	1,000 ft	1 foot	Nonmetallic pipe only Flow velocity: 1–10 ft/s Offered only as a skilled service at this time Bends (the number and degree of bends will limit travel distance)

CIPP: cured-in-place pipe, FRP: fiberglass-reinforced pipe, HDPE: high-density polyethylene, PVC: polyvinyl chloride

of water leaks. A city's entire water pipeline environment can be assessed at once, with the ability to identify leaks within up to a 300-ft radius buffer.

A high-level breakdown of the data capture, analysis, and display process for the satellite-based leak detection technology is enumerated here and shown graphically in Figure 7-11.

1. A microwave sensor acquires images.
2. An algorithm prepares raw data for analysis.
3. A corrected microwave image is analyzed and freshwater leaks identified. The sizes of the leaks are estimated by cross-referencing the algorithm's output against local infrastructure.
4. Normalized data are presented graphically with findings displayed on a web-based application. Field teams on the ground receive "leak sheets" that allow them to confirm and repair the leaks.

Currently, this technology is only available through the services of a skilled and specially equipped service provider. Since satellite-based leak detection scans wide areas of the water distribution system, the utility operator must be ready to respond to potentially large numbers of leak reports submitted. The utility should have a means to prioritize the repair actions for the reported leaks and manage this backlog in a timely manner. Table 7-14 summarizes the capabilities and limitations of satellite-based leak detection.

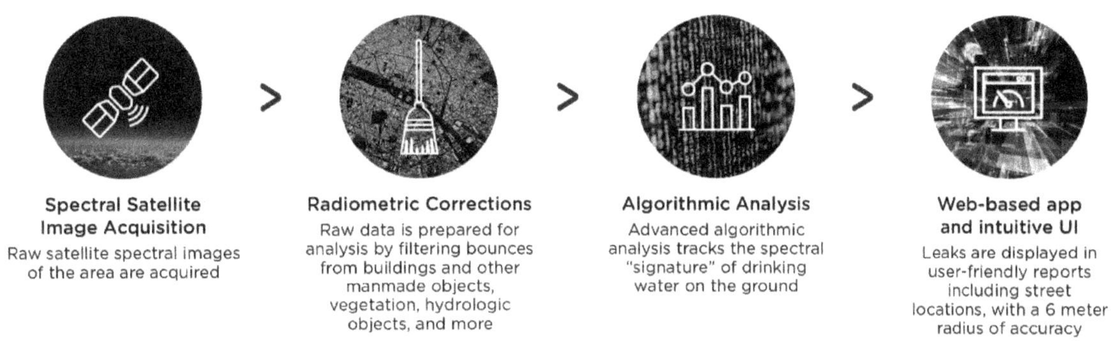

Source: Utilis

Figure 7-11 Steps in data capture, analysis, and display with satellite-based leak detection

Table 7-14 Parameters for satellite-based leak detection

Pipe Materials	Diameter Range	Distance Range	Locational Accuracy	Limitations
All	All	Multiples of 1.50 mi^2	±75 ft	Depth of pipes less than 10 ft Offered only as a skilled service at this time

SUMMARY

While leak detection is primarily performed to identify the sources of real water losses in distribution and transmission systems, the information can also be used to help assess the condition of the pipe and the network. This chapter provides a framework for how information from leak detection can be used in support of condition assessment. Some tools combine leak detection methods with other condition techniques.

At a minimum, the results of leak detection inspections, surveys, and tests can be used as screening tools for prioritizing areas for more intensive inspection with more advanced inspection tools. With recent advances in leak detection technologies, the capability now exists to obtain information about the structural condition of the pipe wall in addition to finding leaks. Considering the relatively low cost of most leak detection technologies, it is prudent to include them as an element of any comprehensive condition assessment program.

Further details on some of these technologies can be found in AWWA M36.

REFERENCES

AWWA (American Water Works Association). 2016 (4th ed.). M36, *Water Audits and Loss Control Programs*. Denver: AWWA.

Sturm, R.; Gasner, K.; Wilson, T.; Preston, S.; and Dickinson, M.A. 2014. Real Loss Component Analysis: A Tool for Economic Water Loss Control. Denver: Water Research Foundation.

Grigg, N. 2017. Visual Guidance for Common Pipe Failures. Denver: Water Research Foundation.

Chapter 8

Internal Remote Visual Inspection

Andi Corrao, Chapter Lead, infrastructureMD, Inc.
Derek Wurst, Black & Veatch Corporation
Noy Phannavong, V&A Consulting Engineers, Inc.
Kris Embry, Hibbard Inshore

Visual inspection of the inside of a water main is helpful in determining the presence or absence of lining and its general condition. Visual inspection may also detect corrosion, cracks, ovality, or other deformities that indicate pipe damage or distress. These methods are also used in determining the location and characteristics of service taps, valves, and other appurtenances.

This chapter introduces various remotely operated technologies, including digital televising, sonar imaging, and laser scanning of both small-diameter and large-diameter pipes. These technologies are deployed on various platforms, enabling inspections in both wet and dry pipes, inside operating mains, or in mains that are temporarily shut down. Direct observation through physical entry (into larger water mains) is covered in Chapter 9.

With the many options available for internal remote visual inspection, the utility professional may wish to consider a phased strategy where less costly basic methods are employed first, and more expensive and advanced methods are employed when additional qualitative and/or quantitative data are required for a more complete and accurate pipe assessment.

INTERNAL VISUAL INSPECTION AS A CONDITION ASSESSMENT TOOL

Knowing the internal condition of a water pipeline is important in projecting the service life of the asset and possible renewal strategies. Internal corrosion can lead to pipeline failures in water mains; therefore, monitoring internal pipeline surfaces for evidence of corrosion is an important aspect of water main condition assessment. Also, the condition of a pipeline's lining and joints provides a good indication of a pipeline's likelihood of failure. In small-diameter pipelines, evidence of tuberculation indicates hydraulic restrictions in the pipe, which can lead to premature pipeline failure, hydraulic deficiencies, and water-quality issues.

Water main linings, such as cement mortar, bitumastic coal tar, epoxy, and polyurethane, provide internal corrosion protection for metallic water pipes of all sizes. A lining that is failing or is compromised can affect the structural integrity of the pipe. The condition of a pipe's lining, including the presence of circumferential or longitudinal cracks, can be observed during visual inspection.

The condition and integrity of the joints is another important pipeline characteristic that can be effectively observed with internal visual inspections. Often, linings at pipe joints are field applied, making them more susceptible to developing defects than areas with shop-applied linings. Joint defects, (including exposed joints and offset joints), evidence of leakage, and evidence of corrosion at joints are defects that can be observed and documented during remote visual inspection. Figure 8-1 shows screen-captured images from a closed-circuit television (CCTV) inspection of a 21-in. water transmission main. Note the lack of lining at the joint and evidence of corrosion. It is not uncommon to see joints on smaller transmission mains in this condition. Poor workmanship or lack of experience often results in field-applied joint mortar that is missing or defective.

Obstructions and other conditions, such as partially closed valves, debris in the pipe, or tuberculation, can affect hydraulic capacity and the long-term operation of the pipe. These items can also be observed and documented through remote visual inspection.

Internal remote visual inspections can also be used to establish baseline conditions for both new and existing pipes. These inspections can be used as screening tools to locate areas for further detailed testing methods and to determine the most appropriate rehabilitation methods as well as the requirements for performing such rehabilitation. The quantity and location of defects can be determined along with dimensional data for estimating purposes.

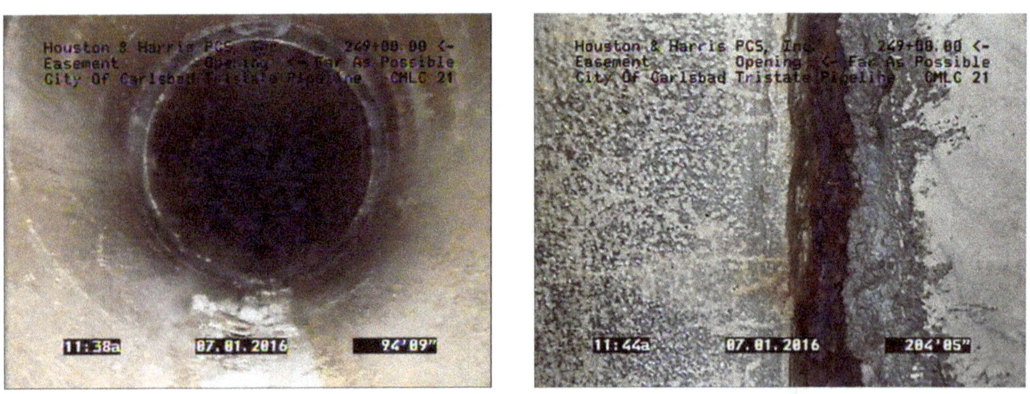

Source: infrastructureMD, Inc.

Figure 8-1 Images from a closed-circuit television (CCTV) inspection of a 21-in. steel water main, with close-up of an exposed joint (right)

RECENT ADVANCES IN REMOTE VISUAL INSPECTION TECHNOLOGY

Constant advancements are being made in remote visual inspection technology. Equipment has become smaller, less expensive, and more capable; the same is true with components/sensors for use with inspection equipment. Remotely operated vehicles (ROVs), including crawlers, swimmers, and floating platforms, have continued to improve with extended reach capabilities, control, communications, and automatic navigation. The main benefits of these improvements translate into better data, less expense, smaller/lighter sensors, and more maneuvering capability for ROVs.

ROVs have advanced from copper wire to fiber optic cable communications. Copper wire limited the travel distance of the traditional ROV to approximately 1,500 ft because of signal attenuation issues; however, the newer fiber optic ROV can operate several thousand feet away from the operator. Battery-powered ROV development, data storage, and battery power capabilities have greatly increased in recent years.

Newer cameras have increased resolution and better zoom and focusing capabilities. Some cameras now use side scan technology that collects images along the sides of the vehicle and ties the images together to create an unfolded and/or virtual walkthrough of the pipe (see Figure 8-2). This technology can expedite the inspection by eliminating the need for stopping and focusing on a defect. High-definition cameras with higher optical zoom capability can allow a better focus and up-close, clear views from greater distances. Advancements in LED lighting have also contributed to better imaging capability due to increased lumens and lower energy consumption. Most cameras are water resistant, and some have significant depth ratings.

Often, water mains cannot be shut down for internal inspections. This has driven the industry to develop digital video inspection platforms that can be deployed in live-flow pressurized pipe conditions. These tethered and untethered devices can travel long distances in water mains while capturing video data and listening for leaks with acoustic sensors. Figure 8-3 shows photos of a video and acoustic sensor head and field deployment of the tethered sensor head into a live-flow water main.

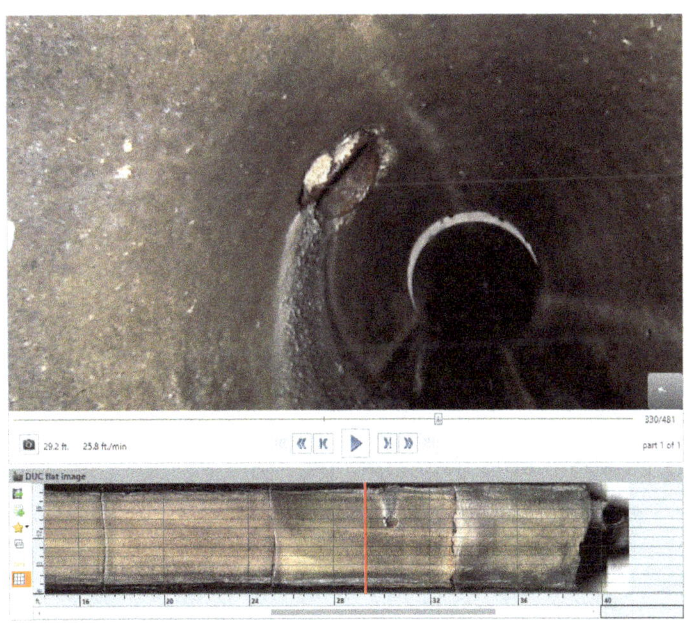

Source: Cues Inc.

Figure 8-2 Example of a side scan camera deliverable

Source: Black & Veatch

Figure 8-3 Digital video and hydrophone sensor head (left); field deployment of tethered sensor head into live transmission main (right)

Although high-definition cameras and advanced sensors have improved the ability to conduct remote internal visual pipeline inspections, physical entry inspection (discussed in Chapter 9) can still provide value to many condition assessment projects. Advancements in portable cameras and nondestructive methods such as ultrasonic testing, surface-penetrating radar, electromagnetics, slab impulse response, rebound hammer, and other tools discussed in Chapter 9 provide an inspector with a multitude of inspection techniques to capture data that complement visual observations.

LOGISTICAL CONSIDERATIONS FOR PERFORMING REMOTE INTERNAL VISUAL INSPECTION

The most appropriate and most beneficial visual inspection method for any given pipeline or group of pipes is dependent on several factors.

Shutdown, Isolation, and Dewatering of Pipeline

One of the most important considerations for performing internal visual inspection of a water transmission main is whether it can be shut down, isolated, and dewatered. This will dictate which method(s) can be practically deployed to inspect the pipeline interior.

Existing Access Points Into Pipeline

Another significant consideration for water mains is whether there are access ports, flanges, manways, or launching stations on the pipeline. Having to install new access points adds considerable cost to the inspection. It is also important to know where the access points are located and how accessible they are. Differing methodologies require different access criteria. For instance, on smaller-diameter distribution piping, digital cameras can now be inserted into pipes as small as 4-in. (10-cm) diameter through fire hydrants. A stiff cable is used to push the camera through the hydrant, into the pipe, and up to 300 ft (100 m) down the pipe in each direction of the hydrant tee. The methodology for using a free-floating platform is quite different and requires both an upstream and downstream access point: one access point for launching and one access point for retrieving the inspection tool.

Pipeline Operating Conditions

For visual inspection of operational pipes, pressure and flow velocity are important considerations. Pressures above 200 psi or flow velocities less than 1 ft/s make the use of most tethered and nontethered devices that are propelled by the flow of water impractical. On the other hand, flows over 10 ft/s can cause visibility issues and possible damage to the inspection equipment.

Constraints Due to Pipeline Configuration

Lastly, the physical configuration of the pipeline (alignment, length, potential obstructions) needs to be considered.

Pipe bends and pipe slope. Changes in direction, both horizontally and vertically, could hinder internal inspections by limiting the distance covered from an access point by either an ROV or a tethered platform. Cable drag increases significantly at pipe bends for digital video cameras. Also, sharp bends can damage the ROV. Generally, only three (maybe four) 90-degree bends can be traversed; otherwise, binding of the cable will prevent the tool from being pulled back.

Pipe length. Long travel distances may make some ROVs impractical. Tethered cameras will always be limited by cable length, as well as by the number of bends and operating conditions (flow and pressure).

Obstructions. Internal obstructions created by valves, appurtenances, debris, and buildup (tubercles) could limit the length of internal inspections with ROV/tethered platforms. For example, butterfly valves and other non–full-bore valves will typically create impassible obstacles for ROVs and tethered sensors due to the risk of getting the equipment or sensor stuck. Cleaning can address the issue of scale and debris buildup, but cleaning requires an additional step and likely an additional shutdown, and there is the potential for complaints of discolored water if the cleaning program is not well managed.

CURRENT METHODS AND TECHNOLOGIES AVAILABLE FOR INTERNAL VISUAL INSPECTION

This section presents visual inspection equipment and platforms currently available for deployment in water mains. First, as a summary, Table 8-1 shows available inspection platforms/methods and sensors for different pipe conditions. Physical entry/diver inspections, listed as available inspection methods in Table 8-1, are presented and discussed in detail in Chapter 9.

Closed-Circuit Television

CCTV inspection is a term commonly used in the industry for what has now become inspection with high-quality digital cameras. Traditionally, these cameras have been video, often remotely controlled, with "pan and tilt" capabilities that allow the operator to zoom in and focus on specific defects. More recently, devices have been introduced that take a series of digital photographs of the entire pipe surface, with software that digitally stitches the photos together, producing a single, continuous image of the interior surface of the pipe.

CCTV cameras generally provide only qualitative visual data, except when using a camera for photogrammetric techniques (dimensioning using photos/video) or with software enhancements (pixel measurements or ranging lights). Additional sensors are added

Table 8-1 Inspection platform/method options

Pipe Status	Available Platforms/Methods	Available Sensors
Dry/dewatered	Physical entry (see Chapter 9) Crawling ROV Push camera/sensor	CCTV Laser GPR UT sensors Thermal camera
Partially flooded	Crawling ROV Floating ROV Free-floating sensor Swimming ROV Push camera/sensor Physical entry/divers (see Chapter 9)	CCTV Laser GPR UT sensors Thermal camera Sonar Acoustic
Flooded or pressurized	Tethered camera/sensor Push camera/sensor Free-floating sensor Crawling ROV Swimming ROV Divers (see Chapter 9)	CCTV Acoustic Underwater laser UT sensors Thermal camera Sonar

CCTV: closed-circuit television, GPR: ground-penetrating radar, ROV: remotely operated vehicle, UT: ultrasonic testing

to the camera platform when the asset owner needs more information or when the conditions warrant it due to turbidity, pipe surface color, access, or pipe size. For example, a standard camera and crawler may not be capable of being submerged or traveling long distances. Or a pipe with black coating may have limited visibility, reducing the usefulness of CCTV data. A submersible crawler or swimming ROV with fiber optics may be the solution for flooded conditions and long-distance travel requirements. Black coatings or surfaces with poor reflectance may require the use of laser or sonar sensors to obtain quantitative data of the pipe wall surfaces; this is preferable to relying on visual information alone (see Figure 8-4). A major challenge with CCTV technology is providing sufficient lighting to get good visibility of the pipe interior surface, particularly for large-diameter water mains (i.e., mains greater than 48 in. [120 cm] in diameter).

Since the visibility of cameras requires either dry pipe conditions or low turbidity in flooded conditions, they can be either replaced or supplemented with laser or sonar sensors. (See the section entitled "Advanced Sensor Technology" later in this chapter.) Technologies such as laser and sonar can often be added to the CCTV camera platform to observe pipe conditions in challenging environments and situations.

CCTV for dewatered pipes. As discussed previously, dewatered pipes can be inspected remotely with crawling ROVs. Common crawling ROVs have a range limit of 800 ft (250 m) to 1,500 ft (450 m), whereas ROVs equipped with fiber optic cables can travel several thousand feet with real-time feedback.

Figure 8-5 contains photos of various ROV platforms with CCTV cameras. Figures 8-6 and 8-7 show samples of images obtained with high-definition cameras in dewatered water mains.

Tethered CCTV for fully operational/pressurized pipes. High-definition cameras can give very good visual inspection data in both distribution and transmission mains. A sensor head containing a high-definition camera is inserted into a pipeline under pressure via existing pipeline features such as air release valve piping (for transmission mains) or a fire hydrant (for distribution pipes). If a suitable entry point is not available, a tap as small as 2-in. (5-cm) in diameter can be made to the pipe to accommodate insertion of the camera equipment.

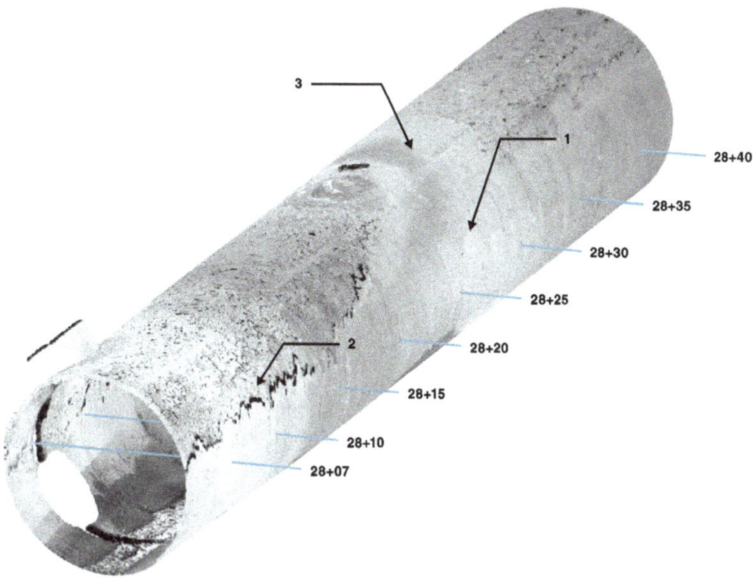

Adapted with permission from Hibbard Inshore

Figure 8-4 Laser scan with better than 2-mm accuracy exported to AutoCAD for manipulation and analysis of areas of concern (i.e., 1, 2, and 3)

Source: Inuktun Services Ltd. and Ibak/Rapidview LLC
ROVs: remotely operated vehicles, CCTV: closed-circuit television

Figure 8-5 Examples of crawling ROVs with CCTV cameras

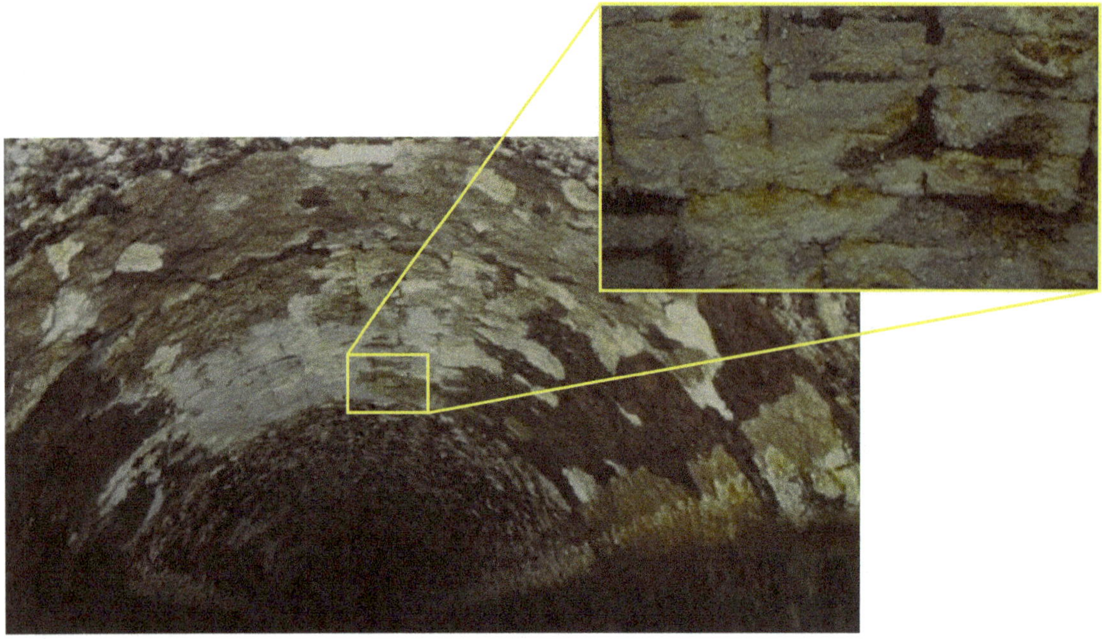

Adapted with permission from Hibbard Inshore

Figure 8-6 Image from a high-definition camera with zoom capability on crawling ROV

Source: Hibbard Inshore

Figure 8-7 Image from a high-definition camera on crawling remotely operated vehicle (ROV) in dry conditions

Table 8-2 Operating parameters for tethered CCTV inspection

Pipe Materials	Diameter Range	Distance Range	Limitations
All	12 in. + for long distance inspection; 4 in. and greater for push cable inspection	Up to 3,000 ft cable length, but generally 2,500 ft max. inspection distance; 250 ft for push cable inspection	Pressure (max. pressure 200 psi; min. pressure 15 psi) Flow velocity (min. 1 ft/s; max. 5 ft/s; not applicable for push cable inspection) Bends (the number and degree of bends could limit travel distance)

CCTV: closed-circuit television

For long-distance travel, a sensor may be propelled by the flow of water using a drag chute that carries the device through the pipeline (see photo showing drag chute in Figure 8-3). The pull on the drag chute caused by the force of the flowing water is controlled by a cable that also carries the video signal data back to the technician. Up to 2,500 ft of pipe may be inspected in this manner, provided only a few bends are encountered along the way. In smaller distribution pipes, a shorter but stiffer cable can be used to manually push the sensor through the pipe for distances of up to 300 ft (100 m); however, the practical distance is about 250 ft (75 m).

These types of inspections are "real time" and, in addition to the CCTV camera, the equipment typically also contains a hydrophone sensor that is used to detect leaks. The tethered system offers the operator increased control over the inspection with the ability to stop the device over particular points of interest and visually assess the condition of the pipe at the time of inspection. Pipe material does not dictate the operability of a tethered system such as this; however, pipe diameter is somewhat restrictive from an operational perspective, as the pipe must be large enough to allow the sensor to flow freely through the pipe. Table 8-2 lists the operating parameters for CCTV inspection in a fully operational water main. Figure 8-8 shows a typical tethered leak detection insertion setup for a long-distance inspection (top left), an insertion through a fire hydrant for a 250-ft inspection (top right), and graphical representation of the tethered operation (bottom).

Additional Platforms for Flooded Conditions

The following platforms can carry a variety of sensors, as listed in Table 8-1 and as described here in further detail.

Swimming ROVs. Self-propelled or swimming ROVs use thrusters to move through a flooded pipe. Capabilities such as speed, maneuverability, and size of the equipment vary widely and are often determined by the size of the access point.

Submersible crawlers (ROVs). A submersible ROV is a self-propelled crawling vehicle that is depth/pressure rated for flooded conditions. Several tractors are available with different size, depth, and range constraints.

Free-swimming multisensor equipment. Neutrally buoyant, free-swimming tools are now available for cable-less video inspection of pipelines with integrated leak detection. These devices use batteries for power, which can limit the range (or duration) of the inspection. Depending on flow velocity and pipeline configuration, inspections of up to 30 mi (50 km) can be achieved from a single deployment. To employ a free-swimming device, the operator must be able to safely and securely launch and retrieve the device and control the direction of the device by closing side outlets. Launch and retrieval can be accomplished using various ports, often without removing the pipe from service.

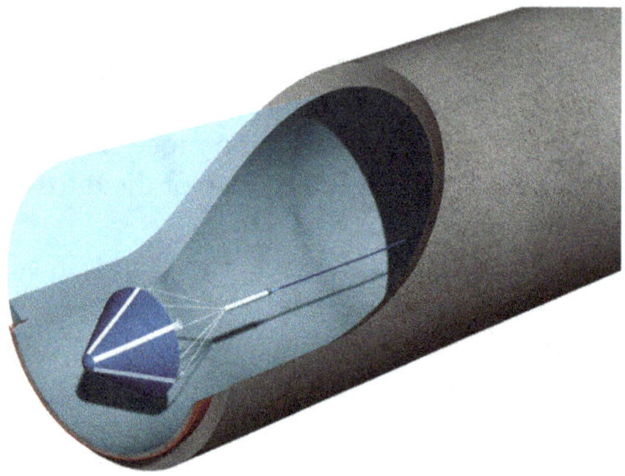

Source: infrastructureMD, Inc., JD7, and Pure Technologies
CCTV: closed-circuit television

Figure 8-8 Tethered CCTV insertion setup (top left), tethered CCTV setup through a fire hydrant (top right), and operation of a long-range system with parachute to propel the sensor with flow (bottom)

Advanced Sensor Technology

Condition assessment using sonar. When internal conditions in flooded pipelines make CCTV inspection impractical or ineffective, "visual" inspection can be accomplished with sonar technology. Sonar uses an acoustic pulse that reflects off of different surfaces, providing a profile of the surface. Sonar offers both qualitative and quantitative information, which can be used in addition to or in place of CCTV in pipes with high turbidity. There are several types and models of sonar equipment. Each has its specific capabilities and constraints. Recent technology improvements have made sonar devices more compact and better designed, with higher accuracy and faster communications than previous models.

Two major categories of sonar commonly used for pipeline inspection are forward-looking sonar devices and profiling sonar devices.

Forward-looking sonar. Forward-looking sonar provides imaging for avoiding obstacles and detecting anomalies such as joints, cracks, offsets, debris, voids (see Figure 8-9).

Profiling sonar. Profiling sonar provides a continuous 360-degree beam to collect data within a pipeline that can detect out-of-round conditions, debris, and other anomalies (see an example in Figure 8-10). One consideration when using or viewing profiling sonar data is that the data are collected in a continuous motion. If the ROV is moving, data may be collected in a helical shape, which may miss some features. Therefore, it is preferable to use a profiling sonar in combination with a camera and/or second sonar.

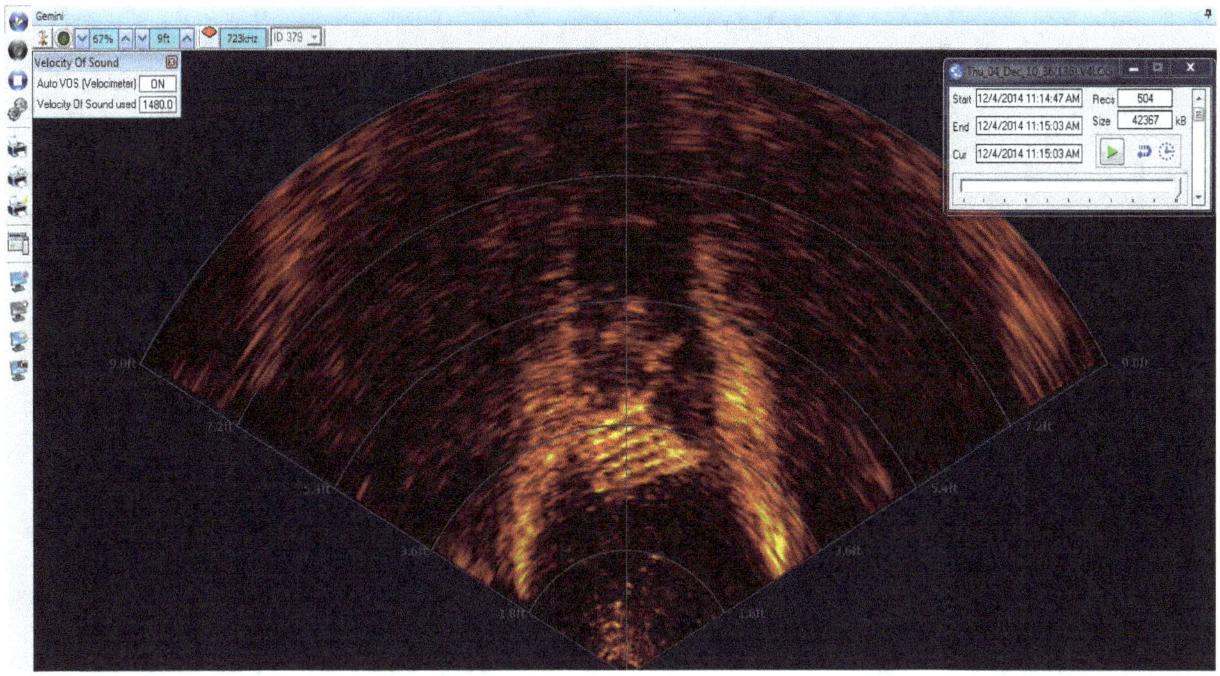

Source: Hibbard Inshore

Figure 8-9 Real-time, forward-looking sonar data showing a grate in a flooded pipe

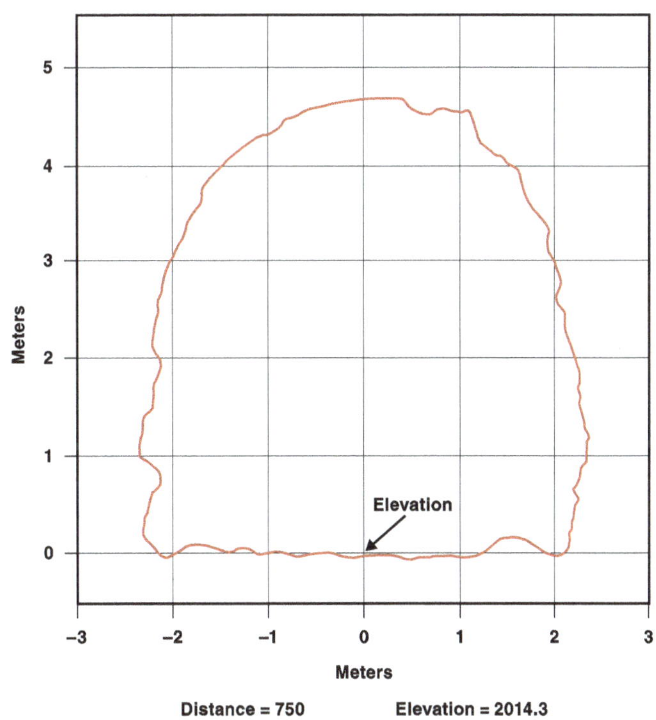

Adapted with permission from Hibbard Inshore

Figure 8-10 Typical cross-sectional view of profiling sonar data

Source: Hibbard Inshore

Figure 8-11 Flooded tunnel visualization using multiple multibeam sonars

Condition assessment using advanced ROVs with multiple sonar sensors. It is possible to collect both qualitative and quantitative data in large-diameter water mains using ROVs with advanced positioning systems. Multiple multibeam sonars can be collected simultaneously to collect pipe/tunnel wall data to show anomalies, restrictions, voids, etc. Figure 8-11 shows an example of a multiple sonar inspection for pipeline visualization in low-visibility conditions. Figure 8-12 shows graphically how the profiling sonar and forward-looking sonar are used together to give detailed information on the interior of the pipeline.

Limitations of sonar. Sonar accuracy is limited because the sound wave expands as it travels outward, and the pulse reflects not only from the target surface but also from other surfaces within the cone-shaped detection zone. Sonar precision is limited because the technology scans a relatively large area from a single point.

Condition assessment using lasers. Laser scanning offers a technological solution to the limitations of sonar and the potential need for high-accuracy spatial information. Lasers typically used in water pipeline applications are ranging or "time of flight" lasers. Ranging lasers measure the time of flight from an emitted light photon to the surface and then reflected back to the laser sensor. Unlike sonar, lasers emit signals that travel in a single direction.

Underwater laser use. Light does not travel through water nearly as well as sound does, but it nonetheless can travel through water in flooded conditions. The limitation of laser use in water is the presence of turbidity, bubbles, and physical impediments. These factors affect the accuracy and utility of laser scans in those conditions.

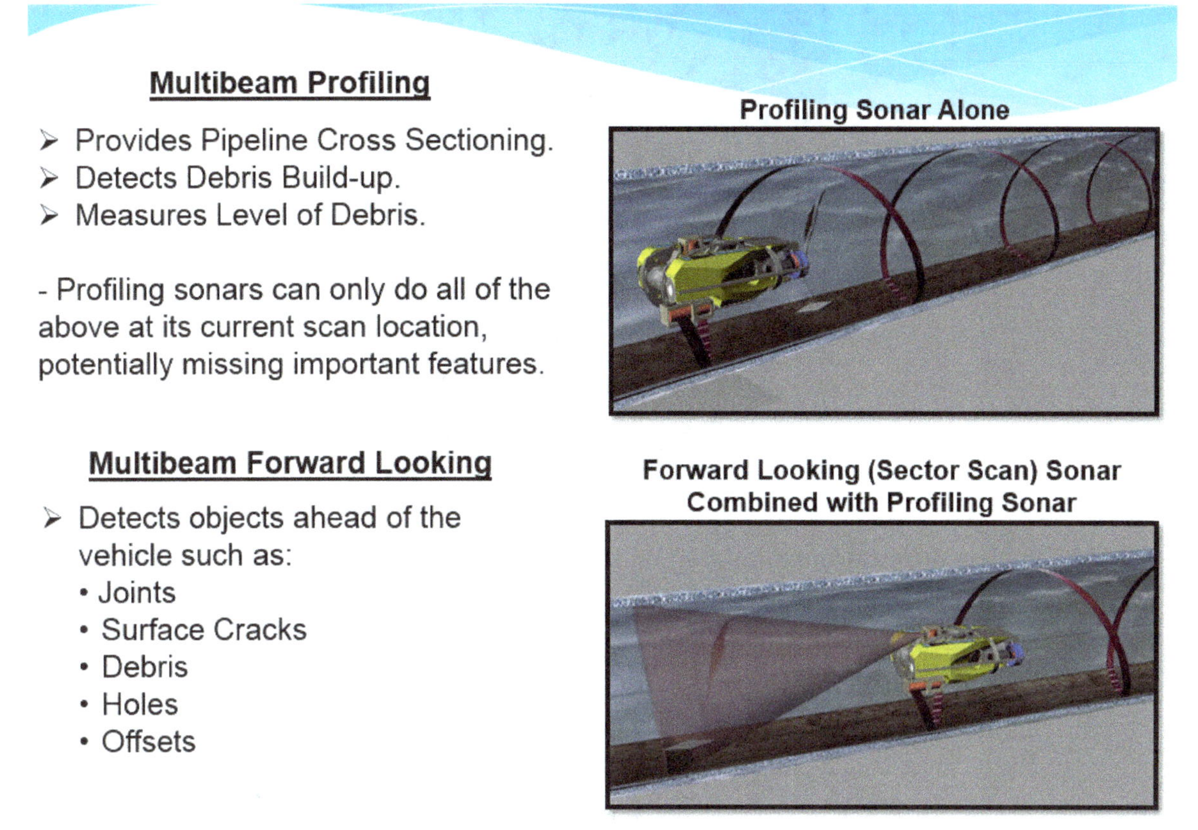

Source: Hibbard Inshore

Figure 8-12 Advantages of combining profiling and forward-looking sonar sensors

With underwater lasers, the sensor head emits a laser line to the target surface, and an optical sensor captures the return signal (similar to sonar). The three-dimensional (3D) positions of hundreds of points along that laser line are calculated, forming a profile of the target. By mechanically rotating the sensor head, a series of adjacent profiles are captured. The sequence of profiles forms a 3D point cloud of the target surface. Laser light emitted from the head of the sensor remains in a narrow band, compared to a sonar beam, which spreads out as it propagates. The energy footprint of a laser beam is therefore far smaller than that of a sonar beam, permitting higher-resolution measurements. Figure 8-13 shows an example of an underwater laser scan.

Above-water laser use. The limitations of laser use above water stem from maximum range, presence of mist, and physical impediments. These factors affect the accuracy and utility of laser scans in those conditions.

Above-water "time of flight" lasers measure the return of light photons that reflect off of a surface. These systems are used in aircraft and automobiles and have been adapted to pipelines. In pipelines, the resolution is affected by pipe size and the deployment method. Therefore, the feasibility and usefulness of above-water lasers is determined by the pipe size, level of detail desired, and available laser configurations and deployment methods. Different laser configurations have different range capabilities and constraints. Some lasers that have been adapted to pipelines can collect data while their inspection platforms are in motion, whereas other lasers require the platform to be stationary (a stationary platform generally produces more accurate data). Figure 8-14 shows a pipe profile collected using a stationary laser platform.

Source: 2G Robotics

Figure 8-13 Sample output from an underwater laser scan of a horseshoe-shaped conduit

Source: Hibbard Inshore

Figure 8-14 A laser scan of a dewatered conduit using a stationary laser platform

IMPLEMENTATION STRATEGIES FOR INTERNAL REMOTE VISUAL INSPECTION

As discussed in this chapter, internal visual inspections can be very useful for assessing the condition of water mains. Engineering judgment of visual inspection data can have considerable value in evaluating the overall condition of a pipeline.

Logistical considerations are important when planning an internal visual inspection of a water main of any size. The various technologies and platforms will have their own specific access requirements. At times, gaining access to the inside of the pipe can add considerable costs to the inspection. Be sure to account for the cost of needed site improvements such as temporary bypassing, shut-down of the pipeline, dewatering, cleaning, disinfection, and/or removal of pipe sections or fittings. Traffic control and permitting may also need to be considered.

When a pipeline is taken out of service for maintenance or repair, regardless of whether it is a scheduled or an emergency shutdown, an internal visual inspection of the pipeline may be beneficial. These opportunistic inspections allow the owner to "piggy-back" on the logistical preparations related to taking the pipeline offline, resulting in substantial cost savings for the assessment.

Methods and Technologies to Start With

If a water main can be shut down with minimal impact to operations and at minimal cost, traditional CCTV crawlers should be considered first for internal visual inspection, given their relatively low cost and high effectiveness. Should additional data be desired, additional sensors can be used on the CCTV crawler platform. Logistical considerations, as described previously, should also be considered and accounted for. Always confirm that any equipment used in potable water mains is dedicated to potable water use. Wastewater pipes are commonly inspected with video equipment, so equipment should be considered contaminated unless proven otherwise.

When to Select the More Advanced Methods and Technologies

When embarking on a pipeline condition assessment project, the first question to ask is, "What is the purpose of the inspection?" Second, "What level of detail and information is really needed?" And third, "What is the size, condition, and location of the access?" The answers to these three questions will help identify which inspection technologies are most appropriate. Budget and timing considerations will also dictate the most appropriate option.

With the wide range of available sensors and ROVs for visual inspection, focusing on the goal of the inspection will help make the best use of your inspection budget. For example, a baseline inspection in a smaller pipe may only warrant the use of a traditional ROV crawler with a standard-definition camera. Similarly, if a pipeline can easily be dewatered and physical entry would be unsafe or impractical, the traditional ROV crawler may be the right choice for the inspection. More sophisticated sensors and ROVs are usually more appropriate when an asset is of high value (or high consequence), has a history of leaks or failures, has difficult access requirements, and/or the level of detail that is required of the inspection data warrants the expense of more advanced technologies.

The need for more advanced technologies for internal visual inspections generally falls into two categories.

1. **Shut-down is not desirable**. The water main cannot be shut down or a significant cost or impact to operations will be incurred during a shut-down of the main. In this situation, it would be beneficial to deploy technologies that are designed to conduct inspections in fully operational pipelines. An added benefit here is that these technologies often come with additional acoustic and ultrasonic sensors that can detect leaks. (Leak detection requires a pressurized pipe.)
2. **Detailed quantitative data are needed**. Where detailed quantitative data are desired or required for engineering analyses, additional technologies/sensors such as sonar profiling or laser profiling can be added to traditional CCTV crawlers to capture and graphically present quantifiable data for condition analysis.

Fortunately, there are many new ROV and sensor technologies being developed and introduced commercially with greater capabilities than were previously available. It is important to first determine the goal of the inspection and what needs to be known. Examples include general pipe condition (e.g., lining or joint condition); hydraulic restrictions, such as tuberculation, debris, or partially closed valves; physical features, such as ovality or cross-sectional measurements of the inside of the pipe; or offsets and profile changes.

Table 8-3 Relative costs of internal inspection methods and technologies

Method/Technology	Cost	Remarks
Out-of-Service Pipes		
Traditional tethered CCTV	$	Lowest cost. Competitive industry. Multiple companies (and usually local companies) offer this service. Basic data delivery.
Advanced CCTV with laser, GPR, UT, or thermal camera sensors	$$	Higher cost. Fewer contractors offer specialized sensors and trained professionals to operate advanced equipment. Mobilization and freight costs add to the price of the inspection. Specialized post-inspection data processing required.
Flooded or Fully Operational Pipes		
Tethered or cable-less (free-floating) CCTV camera with leak detection	$$$	Relatively high cost. Specialized underwater sensors and advanced deployment equipment required. Few companies offer this service. Post-inspection data processing can be time consuming and requires specialized expertise.
ROV CCTV camera with pipe profiling (sonar and/or laser sensors)	$$$$	Highest cost. Collection and analysis of comprehensive and detailed inspection data are expensive. Very few companies own the equipment required for these inspections and can show a successful track record for high-level data collection, data analysis, and data delivery.

CCTV: closed-circuit television, GPR: ground-penetrating radar, ROV: remotely operated vehicle, UT: ultrasonic testing

Relative Costs of Available Inspection Methods and Technologies

The total cost of a visual inspection project is influenced by many factors, including overall project scope and difficulty; cost of equipment required to complete the project (e.g., a laser scanner capable of less than 2-mm accuracy is more costly than a sonar sensor that may produce ±3-cm accuracy in a given range); availability of access roads for trucks and equipment; required ancillary site work (traffic control, valve removal, crane rental, etc.); sophistication of the required ROV (e.g., long distance transport, higher-specification data); contractual requirements for reporting; and administrative tasks.

Relative costs of the methods and technologies presented in this chapter are shown in Table 8-3. These relative costs consider "all things being equal" in terms of the application of the technology. In other words, the relative costs shown are for comparison of the technologies only and do not take into consideration pipe access costs, which can vary greatly between projects. These relative costs do, however, take into account the data analysis costs associated with interpreting and delivering the inspection data.

SUMMARY

Remote visual inspection provides valuable information on the condition of the inside of a water main. Before selecting a remote visual inspection method or technology for a specific project or pipeline, consider the project objectives. Is the objective to predict remaining useful life, prioritize pipes for rehabilitation or replacement, or use the information as a baseline for future inspections as part of a larger program? What type and level of data are needed in order to achieve these objectives? It is prudent to invest in collecting the appropriate type of data and the right amount of data—sufficient for your needs, but not more than is needed.

As discussed in this chapter, pipe access is always a primary consideration for selection of the most appropriate inspection method. Internal visual inspection is often used with other complementary inspection methods (presented in other chapters) to make informed decisions on the integrity, future operation, and rehabilitation of water mains of all sizes. Internal remote visual inspection is just one of the many condition assessment tools available to utility professionals.

This page intentionally blank.

Chapter 9

Physical Entry Inspections

Derek Wurst, Chapter Lead, Black & Veatch
Noy Phannavong, V&A Consulting Engineers, Inc.
Trent Nedens, Ballard Marine Construction

In larger-diameter pipes, direct observations by an experienced engineer or technician can be invaluable. Observations provide the benefits of internal visual inspection by remote methods (as described in Chapter 8), plus the ability to more closely observe and perform physical tests where defects and concerns are observed. Minor repairs may also be performed.

This chapter discusses what can be accomplished through physical entry into transmission pipes. Inspections using people or remote methods are not necessarily mutually exclusive. In some cases, physical entry inspections as described in this chapter and remote visual inspection methods as discussed in Chapter 8 are used together. Both chapters should be read to understand the range of available internal visual inspection methods and their benefits and limitations. Using remote methods in tandem with physical entry inspection can yield a much more comprehensive inspection.

This chapter also describes how physical entries can be performed to obtain visual inspection information on the condition of water mains. Historically, inspections of water transmission mains have frequently employed physical entry. Recent enhancements in physical entry condition assessments have come from the use of various nondestructive inspection tools, as described in other chapters. In fact, many large-diameter inspection tools require physical entry to assemble the device in the pipe and to physically move the rig through the pipe.

Before determining that a physical entry inspection is a viable assessment method, one must consider a number of access-related issues. A range of logistical considerations related to pipe access, including how a person could enter the pipe, will be addressed in this chapter. Safety is always the top concern. There are a number of safety-related issues with physical inspections.

There are many benefits to physical entry. Personal observations and spot testing can be performed very cost-effectively using well-trained and safety-certified professionals and technicians inside the pipe.

OVERVIEW: PHYSICAL ENTRY INSPECTIONS

Effective planning and preparation for a pipeline inspection via physical entry can be extensive, depending on the objectives of the inspection and the physical and operational characteristics of the pipeline. Many of the water pipelines that are large enough for physical entry are also highly critical assets for which outages are difficult to schedule. However, the importance of these highly critical assets also merits the investment in time and resources needed to produce high-quality condition information.

Regardless of the overall objective, executing effective physical entry condition assessments requires experienced inspection personnel with sound knowledge of pipeline design, construction, and repair. They also must be knowledgeable about common deterioration mechanisms and the proper methods to accurately identify and assign a "grade" to associated defects.

An added benefit of physical entry inspection is that identified defects may necessitate the need for repairs, which may also require contractors to physically enter the pipeline. Thus, the value in physical entry pipeline inspection is realized twofold when pipeline owners, operators, and contractors become knowledgeable about the provisions needed to make the pipeline available for entry.

Pipeline Entry Limitations

Condition assessment by physical entry inspection is governed by physical, operational, and safety-related constraints. Constraints may include pipeline diameter, inspection length, access locations, and operational requirements for pipeline segment isolation and dewatering.

A commonly used threshold for pipeline diameters where internal physical entry is feasible is 36 in. (92 cm). Figure 9-1 shows physical entry into various diameter pipelines. Typically, the deciding factor on pipeline diameter is less related to access feasibility than it is to the appropriateness of physical entry inspection versus other suitable nonentry inspection techniques. Pipelines as small as 16 in. have been inspected using physical entry.

Although no standard distance limitations for inspection exist, inspection distances are typically limited by safety considerations for ventilation, communication, and retrieval. Communication between the pipeline entrant(s) and attendant(s) is typically problematic in long-distance inspections. Typically, radios are used for voice communication, but the range can be limited. Air horns have been used to effectively increase the communication range, but they are also limited in their ability to convey anything other than simple messages. For long-range communication, the use of a commercial diver communication box cabled radio is also an option. As with all confined spaces, a retrieval method is needed.

Planning for Physical Entry Inspections

Physical entry inspections are typically divided into two phases: the pre-entry phase and the inspection phase. Pre-entry activities prepare the personnel and pipeline for safe entry

Source: Helix Water District

Figure 9-1 Demonstration of physical entry in various pipe diameters

and inspection. The pre-entry activities include the preparation of a detailed inspection plan—a comprehensive document outlining the objectives of and the steps required to execute the pipeline inspection. The inspection plan should include sections on the purpose and objectives of the inspection, copies of the available record drawings, a discussion of previous studies and maintenance history, a detailed list of pre-entry and inspection activities, and a safety plan.

Although a safety plan should identify potential hazards that may be encountered, actual site conditions may be different than anticipated. In most situations, an inspection follows a plan, and if the plan is written with flexibility, unforeseen field conditions can usually be addressed without difficulty.

Safety considerations. Safety is the highest priority for internal physical inspections. Safety regulations applicable to inspecting pipelines must be addressed during the pre-entry phase and the physical entry inspection. In the United States, the provisions from the Occupational Safety and Health Administration standard 1910.146 (OSHA 2016) applies to the inspection of pipelines.

The safety plan should address, but not be limited to, the following considerations: confined space permitting, pipeline dewatering and ventilation, lockout-tagout (LOTO) procedures, personal protective equipment (PPE), first aid, standby rescue services, and communication and lighting requirements. The safety plan should provide details for emergency extrication procedures and clearly identify key emergency contacts. Physical entry inspection should be carefully planned and executed with extreme discipline. An accident within a pipeline can be fatal.

Primary hazards common to pipeline inspections such as atmospheric hazards and inundation must be considered. These hazards should be identified through a job hazard assessment and must be mitigated before and throughout the inspection. The threat of inundation can be prevented by effective LOTO protocols. Typical LOTO involves both physical and administrative functions. Physical LOTO requires that pumping stations be temporarily disabled and prevented (by locking) from operating. Administrative LOTO requires that a notification tag be attached to the equipment to inform others not to place the equipment back into operation.

Threat from inundation can be mitigated using a "double block and bleed" method. Double block and bleed means the closure of a pipe by closing and locking or tagging two in-line valves and by opening and locking or tagging a drain or blowoff valve open in the line between the two closed valves. This method provides redundancy through double isolation of potential energy sources for the pipeline to be entered. There should always be at least two points of separation of water behind valves or bulkheads so that the entrants would have time to be extracted from the confined space if something were to go wrong.

Most atmospheric hazards can be mitigated by using fans and natural ventilation. The oxygen consumption characteristics of the biomass in the pipe may be unknown. A ventilation calculation can be performed to determine the size of the blowers needed to adequately ventilate the pipeline. Air exchange rates of between 6 and 20 air changes per hour, depending on type of confined space and local regulations, and a consumption rate of 100 ft^3/min per person can be used to calculate cubic feet per minute airflow for the pipeline ventilation system. The ventilation system should also consider inspection team comfort in air velocity. A light breeze in the pipeline can help provide evaporative cooling and humidity control. Air velocities that are too high can increase risks of hypothermia if entrants become wet during the pipeline inspection activities. For entrant comfort, a 1-mph rate is an appropriate air velocity. Using information for ventilation system calculations, the time required to complete an air exchange inside the pipeline can be determined.

An important consideration in setting up ventilation for physical entry are dead ends where the pipe may be flooded or a physical feature may prevent airflow. This can create a zone in which a hazardous atmosphere becomes trapped. It is also important to take caution to prevent sources of airborne contamination from entering the pipeline by keeping the air inlets clear. Exhaust from motor vehicles and generators should be positioned so as to not be drawn into the pipeline during inspection.

In situations where the atmosphere is hazardous, the entrant can be equipped with self-contained breathing apparatus (SCBA). SCBA systems have limited air capacity, so it is important to size the air tank to the length of time the entrant would have to use the SCBA tank before the tank cylinder is depleted. High-pressure air cylinders are commonly available with 60 minutes of air capacity and are constructed of light-weight carbon fiber–wrapped aluminum. Surface-supplied air systems provide much more air capacity, but the length of the entry will be limited to the length of the air hoses. Note that appropriate training is required for personnel to use SCBA or supplied air systems. Training requirements may involve a respirator fit test and a physical spirometry test administered by a medical facility to measure lung function, specifically the amount (volume) and/or speed (flow) of air that can be inhaled and exhaled.

Equipment and personnel transport devices can be helpful in physical entry pipeline inspections to haul inspection equipment and to cover long distances, but most importantly to increase inspector safety and comfort. The transport devices can be simple or custom configured devices. When using transport devices in the pipe, it is important to ensure that they do not become a hindrance if the inspector should need to vacate the pipeline immediately. When properly designed, the transport devices can be used as a platform to carry test equipment into the pipeline, thereby reducing fatigue on the inspector and increasing overall pipeline inspection safety.

Underwater inspections (divers). Divers can enter both semi-submerged and submerged areas with the use of surface-supplied breathing equipment. Depending on the physical constraints of the particular inspection, they can penetrate up to several thousand feet of pipeline or tunnel. The introduction of penetration into an enclosed space by commercial divers increases the risk already posed by any in-water operations. These risks are mitigated with the strict adherence to US Navy dive tables, real-time video feeds, in-water tenders, and pre-dive planning (Naval Sea Systems Command 2008). Some benefits of dive inspection are the ability to move debris or detritus from the path of the diver to aid inspection, accurate physical measurement using handheld instruments, and the ability to collect tactile information in low-visibility environments.

Diver inspections for submerged conditions. In situations where the pipeline cannot be dewatered, physical entry with pipe divers can be performed. When considering the use of trained Association of Dive Contractors International (ADCI)–compliant commercial divers for inspection purposes, there are several physical constraints that must be considered. These include

- anticipated and maximum possible depth to be encountered during the dive (controls the amount of bottom time available to complete the work as well as the necessary safety/backup protocols);
- elevation of the facility (especially when conducting dives in greater than 30 ft of water);
- diameter of the pipeline (minimum 36 in. is preferred, though 30 in. could be possible)
- velocity of the water flowing through the pipeline (4 ft/s is maximum, 2 ft/s or less is ideal);
- physical diving hazards (pumps, traveling screens, valves, pressure differentials, etc.);
- anticipated visibility of the diver and his real-time video equipment;
- target penetration distance (2,000–4,000 ft is the typical maximum range, depending on the number of in-water tenders needed);
- entry/access dimensions (needs to be large enough to physically pass a diver through);
- joints/bends along the penetration route (can increase the number of in-water tenders needed);
- secondary access points (can limit the penetration distance required); and
- air gap between the water surface and the access point (may require tripod/winch setup).

LOTO is an important step in ensuring the safety of the commercial divers in any confined environment. Physical hazards (as itemized above) can be present when diving in pipelines and tunnels, especially when a total outage of flow cannot be facilitated. It is critical to ensure that the dive team has redundant breathing air systems able to be engaged at a moment's notice and real-time video/communications to the surface.

The pace of the inspection in diving situations is controlled by the pipeline's condition, water visibility, and the diver's ability to safely remain on location and perform the work. Pre-dive planning is a critical step in completing any successful penetration dive, which includes backup plans in the event that the supervisor is forced to abort the planned dive. The use of monitors to display the diver's video feed in real time can help the supervisor and other inspection experts to guide the diver toward particular areas of concern.

In some situations, divers and remotely operated vehicles (ROVs) can be used in tandem to complete the work. Divers can facilitate access to an inspection area for the ROV, and divers can also tend ROV umbilical cords in challenging environments.

Visual inspection observations. To assist in determining the various types of observations made during an internal inspection, it is helpful to establish working definitions for classification of observations. Three broad categories are useful in determining if observations are features, anomalies, or defects.

Feature. A feature is any visually identified point or area of interest. These typically include construction joints, access ports, blowoff or air-release valve connections, and service taps. Observations may reveal a defect occurring at features such as leakage at a joint. However, if a feature is identified as not detracting from the pipeline's performance, no further action is required other than to note its condition and location for future reference.

Anomaly. An anomaly is an observation that deviates from the intended pipeline attributes with respect to physical appearance or operation. Anomalies exclude known construction features but would include observations such as cracks, surface deterioration, obstructions, corrosion, or pipeline modifications. To determine if an anomaly is indeed a defect, further investigation or testing may be warranted. If the anomaly is not deemed to affect the pipeline's performance, then no further action is required other than to document its condition and location for future reference.

Defect. A defect observation, therefore, is defined as an anomaly or attribute that detracts or deviates from the pipeline's intended design. Defects may include cracks, deflection or deformation, obstructions, corrosion, erosion or surface deterioration, and leakage. Evaluating the significance of a defect demands a comprehensive structural understanding of how the pipeline was designed and is intended to function.

Note that not all anomalies are defects and not all identified defects necessarily require an action such as repair. In order to judge the relative severity of the defects and to determine the need for action, the National Association of Sewer Service Companies (NASSCO 2015) has established the Pipeline Assessment Certification Program (PACP), which offers guidance to pipeline owners and operators on the threshold defect levels for action for nonpressurized (gravity) pipes. NASSCO in conjunction with the Water Research Centre (WRc) in the United Kingdom has advanced the pipeline inspection industry for the consistent classification of pipeline observations, and this methodology is applicable to inspection of water pipelines. A similar standard for assigning condition ratings specific to pressurized water mains is currently under development through a Water Research Foundation research program. Regardless of the protocol used, physical entry assessments are benefited by following industry standards.

Recording observations (data management). Before conducting a physical entry pipeline inspection, it is recommended that a protocol for consistent data collection is established. The inspection plan should discuss the methods to be used during the physical entry as well as the sequence and frequency of the observations or measurements to be taken. It is important that observations are well defined with consistent location information.

Longitudinal position reference. Typically, the pipe will be marked using stationing numbers at regular intervals beginning at the point of entry and proceeding longitudinally with the pipe as the inspection progresses. The longitudinal reference markings can also be made at other notable connections and features in the piping. Where possible, it is good practice to have the pipe inspection markings reference the original construction drawing stationing. One method to establish stationing marks for a pipeline is to use a handheld distance wheel.

Pipe stationing should be marked on the pipe walls with marking paint (NSF 61 compliant). In certain circumstances, owners may elect to install permanent station markers to assist in future inspections.

Circumferential position reference. Circumferential positions are typically noted as clock positions, i.e., 12:00 at the crown of the pipe. General protocol is that the orientation of the clock positions are referenced facing toward the downstream direction. This may not always be evident in a water pipe; however, one direction should be chosen and used consistently. The clock positions can be measured in the pipeline using the arc length of the pipe internal diameter divided by 12 equidistant positions (30 degrees). Starting at the invert of the pipe (6:00), other pipe clock positions can be measured along the pipe wall. The invert of the pipe is generally self-evident in the field, although a string bubble level can be used to help identify the pipe invert position. The remainder of the pipe clock positions can be marked with the aid of compass dividers, a distance wheel or fiberglass tape measure. If the internal diameter of the pipe does not change during the inspection, a custom-made handheld jig, such as a rod cut to a known cord length, can be used to also mark the arc length for the pipe clock positions. At times, references to the pipe spring line (3:00 and 9:00) may also be facilitated by the longitudinal weld seams of steel pipe sections.

Documentation. Documenting physical entry inspection observations using photographs and videography can significantly enhance a comprehensive water pipeline condition assessment. A tip for taking photos of observations is that photos should generally be taken in pairs. The first photo can be taken with some perspective as to give the owner a point of reference of the observation. A second photo can provide a more detailed, close-up view of the observation. It is prudent practice to provide some object or use a reference scale when taking detailed photos to provide dimensions of the observation. Additionally, the physical entry observations can be recorded using continuous video documentation with personal action camera digital video recorders. These miniature action cameras feature a wide-angle lens that can provide thorough coverage of the pipeline surfaces. However, it is important to note that the broad angle of view can tend to distort the scale and magnitude of the observations. Therefore, it further supports the use of reference scale in collecting video documentation of observations. Further experience indicates that when the action camera is enclosed in a protective case, the sound quality for voice narration of the video can be limited. In these circumstances, a digital voice recorder works well to capture the inspector's notes without the need to use paper forms.

Electronic handheld tablets are also effective means of collecting inspection information, as shown in Figure 9-2. An advantage of using a handheld tablet is that it combines several inspection tools into one device. For instance, the camera, video, pipeline drawings, and data collection forms can be incorporated in one device. Custom field data collection forms can be developed before the inspection to aid in gathering inspection data.

Regardless of the specific equipment used to support the physical entry inspection, it is important to ensure that the inspection gathers enough data to adequately assess pipe conditions. Since variations in condition should be expected, the inspection plan and the inspector should allow for the frequency of measurements to be increased as dictated by condition.

Source: Black & Veatch

Figure 9-2 Electronic data collection during physical entry inspection

Complementing visual observations with physical testing. Physical entry yields high-quality visual inspection information, but it also can allow for additional physical testing and measurements to be performed. This section provides a brief discussion on some of the potential testing that can be accommodated with a physical entry inspection. A more comprehensive discussion of these various techniques is provided in other sections of this manual. This discussion also supplements Chapter 3 on the use of multiple methods to perform condition assessment of water mains.

Depending on the objectives, the physical entry inspection may integrate visual observation with nondestructive testing and/or destructive testing techniques as detailed in Chapter 6—Spot Assessments. Table 9-1 provides an overview of the types of test methods that can be performed during a physical entry inspection, how the tests are performed in the pipeline, and the experience requirement for the test method.

Table 9-1 is not a complete list of the inspection techniques or the requirements of testing, but it provides information demonstrating that physical entry inspection can provide additional opportunities for data collection beyond visual observations. Also see Chapter 6, Spot Assessments. An important factor in determining the type of testing to be conducted during the physical entry inspection is the appropriateness of the technology in determining pipe condition based on material type, logistical considerations, and overall inspection objectives.

Interpretation and Presentation of Results

One of the primary benefits of a physical entry inspection is the ability of the pipeline owner to obtain direct and immediate feedback on the pipeline condition. Testing results may need to be summarized quickly in the field to inform the owner of pipeline conditions.

Table 9-1 Complementary testing techniques performed during physical entry inspection

Test Method	Description and Applicability for Use During Physical Entry	Inspector Experience Requirement
Sounding	Very suitable: Inspector uses handheld chipping/rock hammer.	Qualitative test requires inspector be educated on distinguishing between solid and hollow sounds.
Rebound hammer	Very suitable: Inspector uses handheld spring-loaded testing device. Testing provides indication of compressive strength.	Data interpretation experience and moderate experience to perform testing are required of inspector.
Half-cell potential measurement	Somewhat suitable: Inspector uses handheld testing equipment to take electrical potential measurements over an area.	Testing requires physical electrical contact to pipe reinforcement and experience with evaluating corrosion potential.
Impact echo	Very suitable: Handheld equipment tests concrete pipe wall thickness.	Testing can be executed with some training.
Slab-impulse response	Somewhat suitable: Specialized electronic equipment is used to determine the relative stiffness of concrete and mortar substrates.	Testing requires instruction on equipment use and inspector knowledgeable in data interpretation. Onsite data analysis is required to validate data quality.
Surface-penetrating radar	Somewhat suitable: Handheld equipment is used to scan an area of the pipeline to evaluate concrete for reinforcing steel embedment depth and subsurface defects.	Testing requires instruction on equipment use and inspector knowledgeable in data interpretation. Some limited onsite data analysis and instrument calibration may be required.
Electromagnetic eddy current	Very suitable: Testing provides indication of metallic pipe wall or reinforcing wire defects. Equipment is typically customized for the pipe to be inspected.	Testing requires skilled inspector to ensure proper data are captured. Some onsite data analysis and instrument calibration may be required.
Ultrasonic thickness	Very suitable: Testing provides indication of metallic pipe wall loss. Electronic equipment is handheld.	Testing requires inspector training on instrument and knowledge of where and how to collect data. Some onsite data analysis may be required.
Broadband electromagnetic	Very suitable: Testing provides indication of metallic pipe wall loss. Electronic equipment is handheld.	Testing requires inspector training on instrument and knowledgeable in where and how to collect data. Some limited onsite data analysis may be required.
Core samples	Less suitable: Core samples can provide indication of concrete degradation but extracting samples from within a pipe can be logistically challenging.	Core sampling requires experience in using the equipment, strategically selecting areas for sampling and making repairs. Specific inspector experience with this testing method in pipelines is essential.
pH	Very suitable: Discrete samples of surface mortar provide indication of pipe degradation.	Minimal inspector experience required. Testing requires some care in collecting samples for laboratory analysis.
Pipe ovality	Very suitable: Testing provides information on pipe deflection and uses handheld equipment.	Limited inspector experience required to execute test.
Thermal imaging camera	Very suitable: Thermal cameras are similar to digital cameras and provide temperature variations that may be indicative of subsurface defects.	Qualitative assessment of possible subsurface defects requires minimal experience in assessment of thermal imaging.

Furthermore, field data may need to be used so that appropriate emergency repair decisions can be made. Physical entry inspection can provide the owner with immediate access to pipeline integrity information. To facilitate the interpretation of results, it is advisable for the inspector to input the test data into spreadsheets on an electronic tablet to quickly summarize the results, with links to photos and test data. Spreadsheet macros can be used to automate data entry. Not only does this allow for much of the logged data to be presented quickly, it also makes the information available to transmit electronically.

Pipeline observations can use qualitative ratings such as "excellent," "good," "fair," "poor," or "critical" to describe the pipeline condition. The ratings must consider not just the severity but also the frequency, size, distribution, and location of defects found in the pipeline. One way to help analyze the physical entry inspection information is to plot the data along the pipeline alignment. Spatial distribution of defects/observations is an effective way to convey complex information. The grouping of defects in a cluster analysis may help pinpoint specific areas of deteriorated pipe. The data can also complement consequence of failure in a risk analysis when condition is evaluated with proximity to above-ground features.

Upon completion of the physical entry inspection, it is vital that records be delivered to the owner and that the owner maintain records of the observations from the assessment. A physical entry inspection provides information on the pipe condition at that moment in time. Pipeline systems are dynamic, and the operating and exposure conditions can change over time. Therefore, it is essential that information from the inspection be kept accessible for use in tracking pipe condition for effective pipeline operations and future planning.

SUMMARY

Conducting physical entries of water mains continues to be a traditional and practical method to obtain visual inspection information on interior pipeline condition. Although logistical considerations for physical entries can be extensive, having a well-prepared plan can ensure that the work can be conducted safely and effectively.

Efforts put toward planning for inspection safety should also consider what data to collect, how to record and characterize the observations, and how the data will ultimately be used to make informed judgments on the condition of the pipeline.

There are various benefits associated with performing physical entry inspections. The visual inspection data can be complemented with nondestructive testing techniques. Whereas remote methods tend to be designed to capture specific pipeline data, a physical entry inspection can adapt to the unique conditions of the pipe and conduct a multitude of tests. Visual inspection data, whether obtained by physical entry alone or in combination with other inspection methods, provide valuable information on the condition of the inside of a water main and should be considered as a viable strategy for any water main condition assessment program.

REFERENCES

NASSCO (National Association of Sewer Service Companies). 2015. Pipeline Assessment Certification Program Reference Manual. Marriottsville, Md.: NASSCO.

Naval Sea Systems Command. 2008. US Navy Diving Manual. Washington: Naval Sea Systems Command.

OSHA (Occupational Safety and Health Administration). 2016. Regulations (Standards 29 CFR) Part 1910.146 – Permit-Required Confined Spaces. Washington: United States Department of Labor.

Chapter 10

Acoustic Velocity Testing

Frank Blaha, Chapter Lead, Water Research Foundation
Kevin Laven and Dave Johnston, Echologics, Division of Mueller
Allison Stroebele, Pure Technologies

Acoustic velocity testing for pipeline condition assessment provides information on the average pipe wall thickness loss over the measured length of the pipe. The actual pipe could be generally degraded over its entire length, or the pipe could have significant degradation at only one or two locations. The technique is often viewed as a screening technique to allow a utility to find pipes in generally poor condition. The technology is nonintrusive, noninvasive, and nondestructive in nature and can be used when the main is fully operational with all connecting valves open and all services active.

The resulting data can be used to inform

- asset management programs,
- rehabilitation and replacement decisions,
- before-and-after construction monitoring,
- evaluation of the pipe's structural adequacy,
- estimation of the pipe's useful remaining service life,
- estimation of the pipe's current and future failure rates,
- selection of mains for additional inspection and analysis, and
- asset valuation.

HISTORY

Acoustic velocity (AV) testing has been commercially available for more than 10 years. One of the first comprehensive pilot studies was performed by the National Research Council of Canada in Hamilton, Ontario, in 2004 (Hunaidi et al. 2004). This was a significant milestone in the evolution of this technology because it was one of the first times that the methodology had been applied to and verified on actual in-service infrastructure.

Since then, the technology has been significantly refined. The database that compares physical inspection results with AV testing results (i.e., validations) is growing, and the method is being applied to a wide range of material types and diameters. Significant events and verification activities in the 13-year history of AV testing are shown in Figure 10-1.

2004: The first application in live networks was performed in Hamilton, Ontario, Canada, with validation on cast-iron mains. The verification was successful, and the technology was judged viable (Hunaidi et al. 2004).

2009: A validation project was performed by Battelle on cast-iron pipe in Louisville, Ky., sponsored by the US Environmental Protection Agency (USEPA). The AV technology predicted the pipe to be in good condition, and this correlated with the actual measured metal loss from exhumed samples of the pipe. However, the technology is used to identify *maximum* thickness loss in each hoop of pipe, with these results averaged over the interval tested (which, for this study were 250-foot lengths), as opposed to *average* metal loss reported by USEPA over that same length. These two figures are different, but not incompatible. The average maximum thickness loss should have been a larger number than the average thickness loss, and this was consistent with the quantitated results of this testing (Nestleroth et al. 2014).

2010: A large validation study in Tacoma, Wash., included both asbestos cement and cast-iron water mains. The validations were successful, and the utility judged the technology valuable (Flynn et al. 2014). A project on cast-iron and ductile-iron pipe was managed by Arcadis and included successful validations on both pipe types in Columbus, Ohio (Echologics 2010).

2012: In the Netherlands, a consortium of water utilities joined in a validation study on asbestos cement pipe and cast-iron pipe with the research institute KWR acting as a third-party evaluator. This study confirmed the value of the AV method and identified regional variations in the Young's modulus for asbestos cement pipe to be an important factor for more accurate results (Beuken et al. 2014).

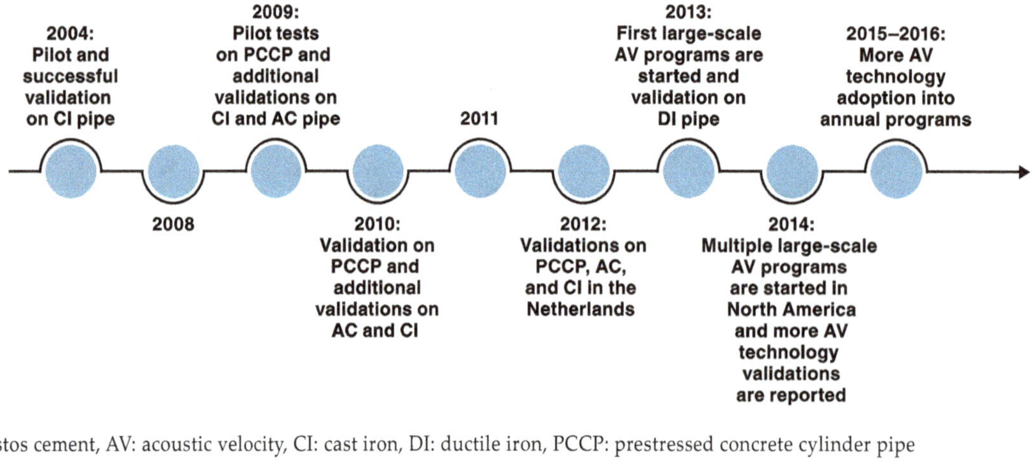

AC: asbestos cement, AV: acoustic velocity, CI: cast iron, DI: ductile iron, PCCP: prestressed concrete cylinder pipe

Figure 10-1 AV testing history and progression timeline

2013: A large-scale cast-iron study using AV technology was implemented by the Washington Suburban Sanitary Commission. The study focused on mains selected for replacement by their asset management computer model, comparing those predictions with the results of condition assessments using AV. Actual condition of the pipe was measured by ultrasonic testing, in-field inspections, soil sampling, and analysis of randomly pulled pipe sections. The AV results indicated that 70 percent of the pipes identified for replacement by the asset management computer model were not in need of replacement and that nearly 40 percent of the pipes identified for replacement by the model were at or near original condition. The results of pipe analysis verified the accuracy of these predictions (Bhimanadhuni and Sinha 2014a). Additional validations were performed by American Water on ductile iron (Bhimanadhuni and Sinha 2014b).

2014: A case study on cast iron was conducted in New Jersey with American Water to differentiate pipes appropriate for structural versus nonstructural lining using AV technology. The AV results were verified by pipe coupon testing at four locations, with the coupon physical testing results matching the AV results (Dasari and Sinha 2015).

2015: In Water Research Foundation Study 4480, relatively good correlations were found between AV tests on asbestos cement mains in the East Bay Municipal Utility District and spot examinations using phenolphthalein stain testing (Spencer et al. 2015).

2015: The total number of validations of the AV method continues to grow, passing 100, and a database is assembled showing a >90 percent correlation between AV predictions and physical testing validations (see Figure 10-2).

BACKGROUND

As pipes degrade, they tend to become less stiff. Acoustic waves travelling along the pipe will be affected by the stiffness of the pipe wall. The velocity of the acoustic wave is slowed where the pipe is less stiff.

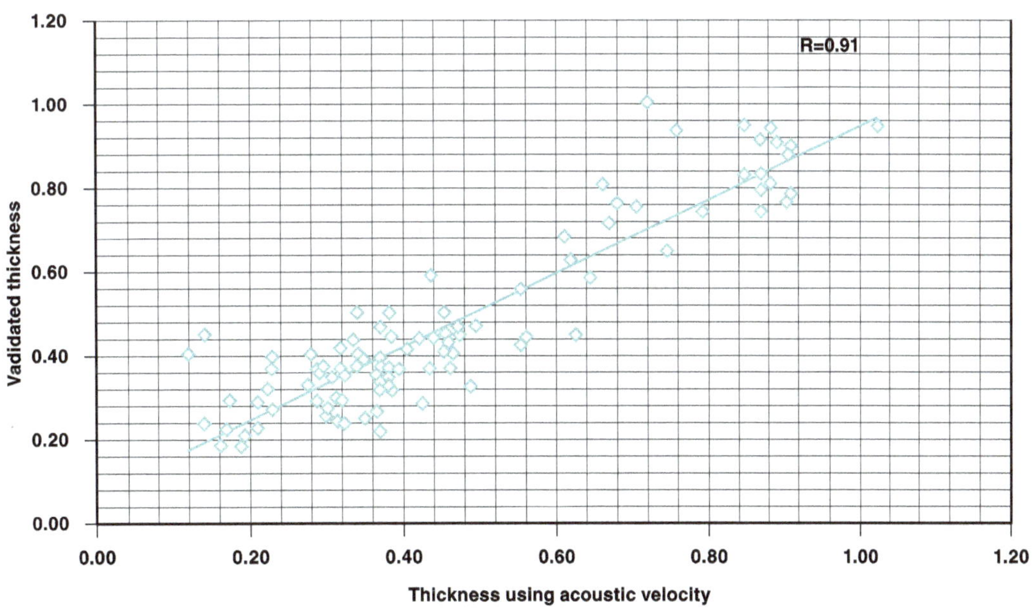

Adapted with permission from Echologics

Figure 10-2 Validation database

For AV technology, an acoustic wave is induced in the pipe either externally or internally. The velocity of the acoustic wave is measured as it travels through a segment of pipeline. The velocity is calculated by measuring the time of travel between sensor locations. Sensors can be positioned externally such as at fire hydrants, or by internal sensors such as hydrophones (USEPA 2014). The velocity (v) can, in turn, be used to calculate the average wall thickness of the segment. To do this, one must know the following information about the main:

- the length of the main between the sensor connection points,
- the internal diameter of the main,
- the pipe material, and
- the water temperature and pressure.

The acoustic velocity will be given by $v = d/t$, where d is the distance between the sensor connection points and t is the time taken for the acoustical signal to propagate between the two sensors. The time taken for the acoustic signal to travel between the sensors is measured using an acoustic correlation tool.

Leak Correlator Method

AV testing is commonly accomplished using leak noise correlators. A typical measurement setup is shown in Figure 10-3. The measurement gives results only for the segment of the main between each sensor connection. In Figure 10-3, this is represented by the span of the main between the valve boxes with the blue transmitters attached.

The average pipe wall thickness between the acoustic sensors is then calculated from a theoretical model. The model describes this basic principle: as the pipe wall thickness decreases over time, the acoustic wave velocity decreases while other variables remain constant or are measurable in the field. The acoustical wave velocity is given in the Wave Velocity Thickness Model as follows, which is based on research from the National Research Council of Canada (Hunaidi et al. 2004).

$$v = v_a \sqrt{1/\left[1 + (D/e)(K_{\text{water}}/E_{\text{pipe}})\right]}$$

Where:

v = propagation velocity of leak noise in pipe
v_a = propagation velocity of sound in an infinite body of water
D = internal diameter of the pipe
e = thickness of the pipe wall
K_{water} = bulk modulus of elasticity of water
E_{pipe} = Young's modulus of elasticity of the pipe material (Hunaidi et al. 2004)

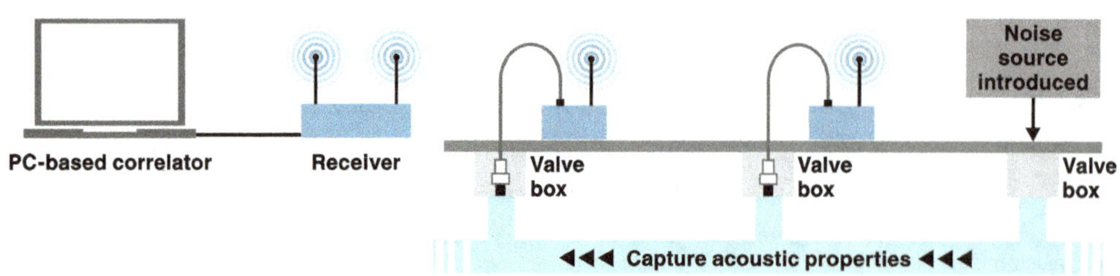

Adapted with permission from Echologics

Figure 10-3 Typical setup for an acoustic velocity measurement

The acoustic wave propagates as a compression wave in the fluid (water), and a dilatational pressure wave in the pipe (the water hammer mode). Water is essentially incompressible in comparison to the stiffness of the pipe wall. Therefore, the pipe must stretch to accommodate the pressure wave. In doing so, the pipe wall goes into stress and then rebounds as the pressure wave passes. The rate of rebound is related to the acoustic velocity and the "structural stiffness" of the pipe.

There are two key implications to this.

1. The structural part of the pipe that can carry load will contribute to the structural stiffness of the pipe. Deposits on the pipe wall such as tuberculation or graphite are much less stiff and will not significantly affect the acoustic velocity and thus are not included in the average wall thickness measurement. Linings, however, do affect results and adjustments must be made—see the section describing linings and coatings later in this chapter.
2. At any given point along the main, the minimum structural thickness of the pipe (where stress and strain are at a maximum) will govern the acoustic velocity. The level of strain of the pipe will be dependent on the minimum wall thickness at any point around the circumference of the pipe.

Therefore, the wall thickness measured with this technology represents an average of the minimum hoop thickness between two sensor connection points. Typically, the length of the pipe segment over which the acoustic velocity is measured is 100–750 ft. For larger-diameter pipe, access may not exist in less than 750 ft. In those cases, excavations using a vacuum truck or other similar device may be needed to obtain acceptable test results.

As noted, the pipe wall thickness calculated from these measurements represents an average value for the pipe segment over which the acoustic velocity is measured. At first glance, this may appear to be a limitation of the technology, as the question of whether this method can find pockets of corrosion could reasonably be asked. However, because the equipment is relatively flexible in regard to test orientation and setup, the survey approach can be changed to accommodate different project goals, and greater resolution can be used to find pockets of corrosion. For example, an asset manager may choose to cover a large area quickly by testing average segment lengths of 600 ft. Following the survey, the data can be reviewed for anomalous results, and segments can be revisited with shorter test lengths to further isolate the location of anomalous results. This approach allows for localization of degraded pipe to within 600 ft, followed by isolation of degraded pipe containing concentrated corrosion pockets to within 100 ft (minimum testing distance depends on access, pipe type, and diameter). For more precise assessment, other methods (as described in other chapters) can then be employed.

Pressure Wave Analysis

A variation of the AV method, which has been under development in Australia for more than 17 years, has recently been introduced to the North American market. This platform is based on the same principles of AV testing and is applicable to all pressurized metallic, concrete, and asbestos cement (AC) pipes. A controlled pressure wave, which is another term for a high-intensity, short-duration acoustic wave, is introduced into the pipeline and monitored by sensors installed on existing pipe fittings. The transient wave experiences partial reflection when it encounters any change in pipeline structure, and the response of the transient wave to the pipeline system is analyzed to determine anomalies, including air pockets, blockages, and the sealing status of valves and unknown pipe features. Average wall thickness can be measured between two test points, and subsectional average wall thickness can be provided in 30-foot increments over the tested pipe length (Hydromax-USA 2017).

In-Pipe Acoustic Tools

Another platform under development for data collection is an internal system that combines AV testing with leak detection. The system works as described previously, by measuring the acoustic velocity of a wave traveling along the pipe. Pulsers and hydrophones in an array inside the pipe measure the velocity over shorter distances, and the system is equipped with video and leak detection capabilities (Figure 10-4). The sensor array is inserted into an operational and pressurized pipeline. Testing of the technology has shown promising results. Several field trials on AC pipe produce repeatable results where pipe joints, appurtenances, repairs, and material changes are detectable (changes in stiffness). Several pipe sections indicating reduced structural stiffness have been identified, and validations for these results were pending at the time of writing.

PERFORMANCE

The external (leak correlator) technology is considered to be fully developed. As shown in Figure 10-2, hundreds of acoustic velocity tests have been verified using other thickness measurement techniques. These thickness measurement techniques include: coupon sample analysis (Chapter 6), magnetic flux leakage testing (Chapter 12), broadband electromagnetic testing (Chapter 11), phenolphthalein dye testing (Chapter 6), and tensile testing (Chapter 6). The sample set includes cast iron, ductile iron, and asbestos cement. The pipes originate from Washington, D.C.; Ohio; Maryland; New Jersey; California; Kentucky; Minnesota; the Netherlands; Canada; Australia; and Belgium.

As can be seen in Figure 10-2, there is a strong correlation between results measured with other technologies ("Validated Thickness" on the y-axis) and results measured using the acoustic velocity method (the x-axis). The R value (Pearson correlation coefficient) is 0.95, indicating a strong linear correlation between the data sets.

This is an exceptionally strong correlation considering that other measurement techniques have different principles of operation and their own inherent errors. There are two factors that usually contribute to some expected errors between the test methods, including the following.

1. Other thickness measurement techniques vary across the data set. Each method has its own limitations and errors, which compounds errors on the data set.

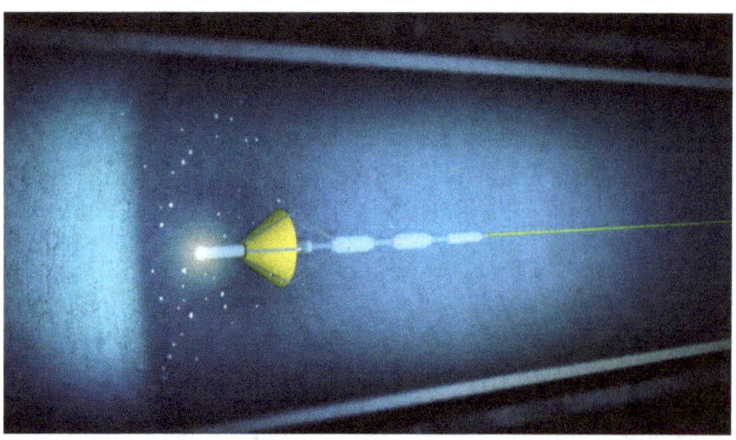

Source: Pure Technologies

Figure 10-4 In-pipe acoustic velocity system

2. The AV method measures an average thickness between two sensor connection points (a pipe segment spanning hundreds of feet in length). The validation methods were applied to segments of pipe five feet or less in length. Since it is expected that pipe condition varies along the main, it is improbable that a short sample of pipe would provide a representative measurement for the entire length. Consequently, increased error would be expected in the data set.

The validation testing results that are available are positive, and several different utilities and organizations have had good experiences applying this technology. Still, more robust validation testing, especially third-party, independent testing (working closely with the vendors) would help to advance the acceptance and application of this technology.

Applications, materials, and limitations of AV are discussed in the following sections.

Uses on Different Materials

Cast iron. The AV method is applicable for use on pit cast and spun cast iron. This method provides pipe wall condition results independent of the existence of tuberculation and graphitic corrosion, which are common ailments of cast-iron mains. For both types of cast-iron pipes, AV testing measures only the structural part of the pipe wall. Softer materials such as tuberculation and graphitic corrosion do not contribute to the structural wall thickness and therefore are not measured using this method. Some pit cast and some spun cast-iron pipes were installed with a cement-mortar lining, and some of these pipes may have been cement mortar lined since installation. Similarly, some of these pipes may have received a polymeric lining since installation. In the instances where the pipe has a lining, adjustments are needed to account for the structural contribution of the lining to the acoustic velocity. These adjustments are described below in the linings and coatings section.

Asbestos cement. AC pipe degrades very differently than metallic pipes. As AC pipe ages, there is gradual leaching of calcium out of the pipe wall, particularly where water is soft or soils are acidic. As leaching progresses through the pipe wall, it leaves behind a soft material that has less strength. AV testing only measures the structural part of the pipe wall and thus is a useful technology for assessing the condition of AC pipe.

Because AC pipe often degrades more uniformly than iron pipe, it may be well suited for AV analysis. In applying AV technology to AC pipes, however, some considerations were identified in a recent study that can impact the interpretation of AV testing results. First, in this case it was found that the installed AC pipe wall thickness varied significantly from pipe to pipe, which was unexpected and would significantly impact AV results, where a single given original pipe wall thickness is assumed in the evaluation. Also, in comparing AV results to AC pipe stain testing, it was found that inner wall stain loss varied considerably around the circumference of the pipe. Therefore, stain tests performed on coupons, which measure that stain at a single point along the circumference of the pipe, do not necessarily represent average pipe conditions as should be measured by AV methods (Spencer et al. 2015).

Ductile iron. Ductile iron (DI) degrades in a similar fashion to cast-iron pipe, and therefore AV testing is suitable for assessing its condition. Typically, DI pipe is manufactured with a cement-mortar lining, or it may have received a cement-mortar lining or a polymeric lining after installation. In these cases, adjustments are needed to account for the structural contribution of the lining to the acoustic velocity. These adjustments are described in the upcoming linings and coatings section.

Steel. AV testing is applicable to steel pipe because it degrades in a similar manner to other metallic pipe materials, but it also has some limitations in this application. Large pipes with very thin pipe walls have less sensitivity to the AV model. This limitation starts to become particularly apparent when testing steel mains because they are typically large in diameter (greater than 24 in.) and have relatively thin pipe walls by design. The acoustic velocity method is most applicable to steel mains when the diameter-to-thickness ratio is <50.

Concrete-composite. Pipes whose construction includes reinforcing bars, wires, or mesh are considered in this category. They have complex construction, resulting in complex degradation and failure mechanisms. Since AV testing is a direct measure of structural stiffness, it can be used to test concrete-composite pipes for structural integrity. However, there are some limitations to the use of AV technology on concrete-composite mains:

1. The pipe does not lose "structural thickness" as it degrades. Multiple degradation mechanisms typically occur simultaneously to both the concrete and steel portions (steel cylinder, reinforcement, or mesh) of the pipe wall. The result of these complex degradation mechanisms is a loss of structural stiffness that resists both internal and external loads. Therefore, when using the AV method on these pipes, one must be sure to compute the stiffness of the main while holding the thickness constant. This provides a more understandable measure of the pipe condition.
2. The original stiffness of concrete-composite mains is highly variable even when new. Stiffness alone is not a key design criterion and thus it is not closely controlled in manufacturing. Therefore, it is difficult to evaluate the measured stiffness of a pipe and compare it to an original stiffness.

These two factors limit the application of AV testing to concrete-composite pipes, where it can be used as a screening tool before using other technologies.

Linings and Coatings

Independent direct measurement of lining condition is not possible. Lining is accounted for by including it as a contributor to the overall structural stiffness of the main. As liners (and the pipe wall) degrade, the overall stiffness decreases. In turn, this will cause a decrease in the acoustic wave velocity, and the structural thickness can be measured. Ideally, the utility should know with some certainty whether a pipe is lined as this significantly affects the evaluation of the pipe.

REQUIREMENTS AND LIMITATIONS

External System Pipe Access Requirements

For a typical AV test setup, access to the pipe is required at three locations. Two of the locations are sensor connection points, and the third location is a noise source introduction point.

Suitable sensor connection points include

- line valve flanges, bodies, and operating nuts;
- connecting valve flanges, bodies, and operating nuts—the valve must be within less than 0.05 percent of the sensor-to-sensor spacing distance away from the subject pipe;
- air valves, blow off valves, pressure relief valves, and drain valves if they are within less than 0.05 percent of the sensor-to-sensor spacing distance away from the subject pipe; and
- the pipe surface, through small excavations (which can be created using hydro- or vacuum-excavation).

Suitable noise source induction points can be anywhere on or off the subject main, beyond one of the sensors and not further than the test segment length. It is preferred to use fittings where water can be flowed. The water flow can be used as a noise source, and the fitting can be used to measure the water temperature and pressure, both requirements for AV testing. Typical flow rates do not exceed 5 gpm for 5 min per AV test.

Other Limitations and Considerations

- **Low pressure:** Pipes with pressure less than 15 psi should not be tested using the AV method. Low pressure decouples the pressure wave from the pipe wall.
- **Air pockets:** Air pockets act as vibration attenuators, which sometimes stop acoustic waves from travelling down the pipe. Typically, this is only a concern when pipes have recently been dewatered.
- **Repair segments:** Unknown repair segments, especially plastic repair segments, on a pipe of different material will result in inaccurate results.
- **Other interference:** Other mechanical devices such as pressure relief valves and pumps can create significant amounts of vibration energy. AV testing will not work when in the direct vicinity of very loud mechanical devices. It is recommended that AV tests are at least 500 ft away from high-energy noise sources.
- **The fluid:** The pipe must be carrying water. The AV method is not applicable to pipes carrying sewage, oil, or gas.
- **Distance measurements:** The accuracy of an AV measurement is dependent on accurate measurement of the linear pipe distance between each sensor connection point. If there are multiple bends in the main between sensor points, accuracy of the resulting velocity measurement is decreased. Therefore, care must be taken when planning a test setup to ensure that sensor-to-sensor pipe spans have a low number of bends or to place sensors at all pipe bends so that the distance can be accurately measured.
- **Pipeline components:** The characteristics of every component of the system between the two sensors must be understood. These include the lengths, diameters, and materials of each item along which the acoustic wave travels. This is particularly important when the sensors are not installed directly on the main but are on a hydrant or other lateral.
- **Appurtenances**: In general, pipe appurtenances (valves, outlets, joints) have negligible effects on measurement accuracy in AV testing. Most appurtenances do not span a significant linear length of the pipe between sensor connection points, so the contribution from these appurtenances is not significant.

REFERENCES

Beuken, R.; Horst, P.; Diemel, R.; and Mesman, G. 2014. Main Condition Assessment by Echopulse, A Validation of Results. 16th Conference on Water Distribution System Analysis (WDSA), American Society of Civil Engineers, Environmental Water and Research Institute, Bari, Italy.

Bhimanadhuni, S. and Sinha, S. 2014a. Acoustic Condition Assessment of Critical Transmission Main in Barrier Islands Post-Sandy, New Jersey American Water. Sustainable Water Infrastructure Management, Water *Infrastructure* Database (WATER*i*D) and Pipeline *Infrastructure* Database (PIPE*i*D). Blacksburg, Va.: Virginia Polytechnic Institute and State University.

Bhimanadhuni, S. and Sinha, S. 2014b. Pilot Acoustic Condition Assessment of Water Distribution System, Washington Suburban Sanitary Commission (WSSC), Maryland. Sustainable Water Infrastructure Management, Water *Infrastructure* Database (WATER*i*D) and Pipeline

Infrastructure Database (PIPE*i*D). Blacksburg, Va.: Virginia Polytechnic Institute and State University.

Dasari, B. and Sinha, S. 2015. Rehabilitation of Cast Iron Mains, Millburn-Maplewood, New Jersey. Sustainable Water Infrastructure Management, Water *Infrastructure* Database (WATER*i*D) and Pipeline *Infrastructure* Database (PIPE*i*D). Blacksburg, Va.: Virginia Polytechnic Institute and State University.

Echologics. 2010. Acoustic-Based Condition Assessment Provides Accurate Remaining Pipe Wall Thickness Measurements for Columbus, Ohio. White Paper. Toronto: Echologics.

Flynn, R.M.; Coleman, M.; and Laven, K. 2014. Tacoma's Pipeline Assessment Project: Replacing the Right Mains at the Right Time. American Society of Civil Engineers, Pipelines 2014 Conference, Portland, Oregon.

Hunaidi, O.; Wang, A.; Bracken, M.; Gambino, T.; and Fricke, C. 2004. Acoustic Methods for Locating Leaks in Municipal Water Pipe Networks. International Conference on Water Demand Management, Dead Sea, Jordan.

HydromaxUSA. 2017. Pipeline Condition Assessment Technology. https://www.hydromaxusa.com/pipeline-condition-assessment-technology (accessed Feb. 1, 2017).

Nestleroth, B.; Flamberg, S.; Condit, W.; Matthews, J.; Wang, L.; and Chen, A. 2014. Field Demonstration of Innovative Condition Assessment Technologies for Water Mains: Leak Detection and Location. Washington: US Environmental Protection Agency.

Spencer, D.; Ellison, D.; Bell, G.; Reiber, S.; Von Aspern, K.; and Snoeyink, V. 2015. Development of an Effective Management Strategy for Asbestos Cement Pipe. Denver: Water Research Foundation.

US Environmental Protection Agency (USEPA). 2014. Field Demonstration of Innovative Condition Assessment Technologies for Water Mains: Acoustic Pipe Wall Assessment, Internal Inspection, and External Inspection. EPA/600/R-14/148. July 2014.

Chapter 11

Electromagnetic Testing Technologies

Ricardo R. Hernandez, Chapter Lead, Metropolitan Water District of Southern California
Dave Spencer, HDR
Chris Garett, PICA
Joanna Line, City of Calgary
Allison Stroebele, Pure Technologies
Martin Roubal, Rock Solid Group
Rod Jackson, CH2M Hill (now Jacobs)

Electromagnetic (EM) technology has a long history in pipeline assessment dating back several decades. EM technology can broadly be categorized as time domain electromagnetics and frequency domain electromagnetics. Both variations of EM technology are used for pipeline assessment today and are offered as broadband electromagnetics (BEM), a derivative of pulsed eddy current (PEC) and remote field testing (RFT), also referred to as remote-field eddy current (RFEC) or remote field electromagnetic technique.

EM technology can be used to assess the condition of the pipe wall by measuring the relative pipe wall thickness and identifying areas of wall loss and corrosion. The data provided by EM technology can be used to estimate remaining useful life and help inform capital planning decisions to monitor, repair, or replace existing pipelines.

Several commercially available tools have successfully used EM technology; however, there are certain limitations that should be considered before implementation.

The applicability of EM technology to water mains can be summarized as follows.

- Types of applicable materials
  - cast-iron pipe (CIP)

  - ductile-iron pipe (DIP)
  - steel pipe
  - bar-wrapped concrete cylinder pipe (BWP)
  - prestressed concrete cylinder pipe (PCCP) (see Chapter 13)
- Typical pipe diameter ranges: 3 in. and greater
- Typical lining and coating ranges: ½- to 1-in. maximum
- Typical output parameter: wall thickness (with respect to a "nominal" pipe reading)

This chapter addresses

- the history and current use of electromagnetic testing in water mains, how it works, and the current state of technology development and
- what can be detected using this technology, the limits of detection, and what factors affect the resolution.

HISTORY AND CURRENT USE OF EM TESTING IN WATER MAINS

The initial development of EM technology for nondestructive testing began more than 40 years before it was ever used in the water industry. W.R. McLean patented the remote field effect in 1951, and a division of Royal Dutch Shell purchased that patent to develop tools for the nondestructive inspection of casing pipe. In 1961, Thomas R. Schmidt (of Shell) was credited with running the first "smart pig" through a pipeline. Between the 1960s and 1990s, both PEC and RFEC technologies were improved upon and used in many applications, including gas transmission lines, down hole well casings, and in heat exchangers.

The first RFEC inspection of water pipelines took place in 1991, involving CH2M Hill, the City of Edmonton, Alberta, and Russell NDE Systems. Russell NDE Systems had been developing the RFEC technology for use in boiler and tube inspections for industrial clients when they were approached to test the technology on 6-in. cast-iron and ductile-iron pipes that had been removed from Edmonton's water system. The Water Research Foundation (WRF, formerly AWWARF) publicized these early applications as part of Project 507 (Jackson et al. 1992). More recently, WRF Project 4471 and other studies have affirmed the accuracy of RFT results (Ellison et al. 2018).

Since the first successful tests in 1991, the use of RFT inspections in water mains has grown steadily. Today's tools vary in their inspection resolution, typically balancing the tradeoffs between improved data quality and difficulty of pipeline access. Vendors from around the world offer an array of tools for different pipe sizes and applications, including wastewater pipelines.

The first PEC inspection of water/wastewater pipelines took place in 1998, involving Brisbane Water and the Castlemaine Perkins Brewery in the City of Brisbane, Australia, and Rock Solid Pty Ltd. Rock Solid had been developing a technology based on PEC called BEM for use in pipeline inspections for water and sewer authorities (services commonly combined in Australia). Rock Solid tested the technology on an 18-in. mild steel pressure sewer line system. In 2000, the Commonwealth Scientific and Industrial Research Organization (CSIRO) was funded and engaged by a number of Australian water authorities to provide independent evaluation of technology. A similar technology evaluation was undertaken by Gas Technology Institute (GTI) in Chicago.

Since the successful inspection for Castlemaine Perkins Brewing and confirmation of technology applicability by both CSIRO and GTI, the use of BEM inspections in water mains has grown steadily.

Today's tools vary in their inspection resolution, typically balancing the trade-offs between improved data quality and speed of scanning. Vendors from around the world offer an array of tools for different pipe sizes and applications, including wastewater pipelines.

Figures 11-1 through 11-6 illustrate the types of EM technology tools that are currently in use.

Source: Pure Technologies

Figure 11-1 Example of a free-swimming tool

Source: Pure Technologies

Figure 11-2 Example of a high-resolution, large-diameter tool

Source: Pure Technologies

Figure 11-3 Example of a self-propelled crawler

Source: Rock Solid Group

Figure 11-4 Example of a handheld tool

Source: Rock Solid Group

Figure 11-5 Example of a manually winched tool

Source: PICA

Figure 11-6 Example of a hydrant-launched, water-propelled tool

TECHNOLOGY OVERVIEW

Remote Field Technology (RFT)

RFT has nearly equal sensitivities of detection at both the inner and outer surfaces of a ferromagnetic tube.

RFT can be used internally on a pipe to detect internal and external flaws and lends itself extremely well to the detection and sizing of both general wall thinning and localized pitting in cast-iron, ductile-iron, and steel pipes and tubes. RFT does not require intimate contact with the pipe wall, making it suitable for pipes with mortar lining or internal scaling.

RFT will detect and determine the size of graphitization and can detect inclusions and fractures of ferrous infrastructure that may occur within the wall of the metal under inspection.

Some RFT tools can be used to inspect pipes from the outside through external coatings, insulation, and cement mortar. The external coating is ideally less than 3 in. thick and can be covered by a nonmagnetic metallic cladding (like stainless steel or aluminum).

Basic RFT probes (Figure 11-7) have one exciter coil and one detector coil. Both coils are wound coaxially and are separated by a distance greater than twice the tube diameter. This axial distance is characteristic of RFT. If the exciter and detector were to be placed close together, the detector would only be sensitive to the direct energy coupling from the exciter in its vicinity.

The exciter coil is driven with a relatively low frequency sinusoidal current to produce an EM field. This changing field induces strong circumferential eddy currents that extend axially as well as radially in the tube wall. These eddy currents, in turn, produce their own EM field that is picked up by the receiver coils array. By monitoring the consistency of the voltage induced in the pickup coils, one can monitor changes in the pipe wall. The strength of the magnetic field at this distance from the exciter coil is fairly weak, but it is sensitive to changes in the pipe wall from the inner diameter to the outer diameter. The resolution of the tool depends upon the number of detectors used. As many as eight detectors may be used on a 6-in. pipe.

EM tools are introduced to a pipeline at an open end or "launch barrel" and can be pushed by product flow, water, hand, or air pressure to the target distance. They are removed at the far end via a "receiver barrel" or can be winched back using a simple hand or electric-winch. For 6-in. and 8-in. mains, tools may be launched from fire hydrants. Additional information is provided in Table 11-1.

Broadband electromagnetics (BEM). BEM is able to detect flaws on both inner and outer surfaces as well as inclusions, graphitization, and fractures of ferrous infrastructure, which may occur within the wall of the medium under inspection.

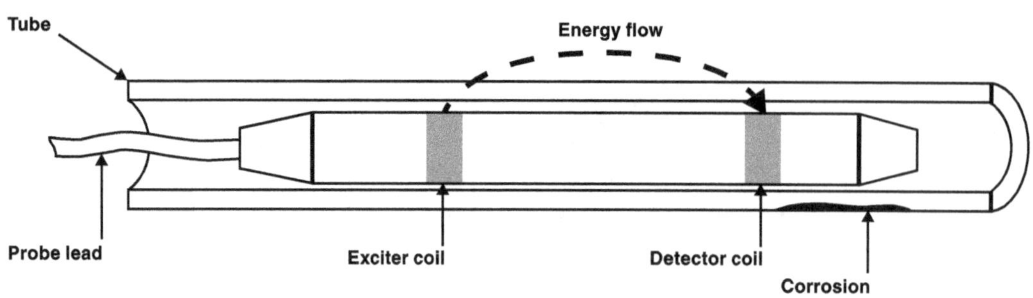

Adapted with permission from PICA

Figure 11-7 Simple probe for remote field testing

Table 11-1 Available tools using electromagnetic technology

	Free-Swimming Deployment		Tethered Deployment		Manual Deployment	
	Neutrally Buoyant Free-swimming	**High-resolution Free-swimming**	**Robotic Crawler**	**Tethered In-line Inspection Tools**	**Handheld Scanner**	**Manned Cart**
Pipe Wall Intimacy	Detectors on stabilizer pedals	Rigid or flexible detector array in close proximity to the inside pipe wall detectors are not in contact with the pipe wall. All contact with the pipe is through soft rubber components to prevent damage to internal cement liners	Rigid array keeps sensors in close proximity to the pipe wall	Rigid or flexible detector array in close proximity to the inside pipe wall detectors are not in contact with the pipe wall. All contact with the pipe is through soft rubber components to prevent damage to internal cement liners	Rigid array keeps sensors in close proximity to the pipe wall	Rigid array keeps sensors in close proximity to the pipe wall
Material	CIP, DIP, steel, BWP	CIP, DIP, steel, BWP	CIP, DIP, steel, BWP	CIP, DIP, steel, BWP	CIP, DIP, steel, wrought iron, rivets	CIP, DIP, steel, wrought iron
Pipe Diameter	16–48 in.	3–36 in.	24–48 in.	3–48 in.	No limitations for external applications Physical entry (Ch. 9) required for internal applications	Physical entry (Ch. 9) required for internal applications
Linings and Coating	Maximum liner thickness of ½ in.	Recommended maximum liner thickness ranges between ½ and 1 in. Thicker liner will impact results	Maximum liner thickness of ½ in.	Recommended maximum liner thickness ranges between ½ and 1 in. Thicker liner will impact result	No limitations	No limitations
Outage Required	No	During insertion	During insertion and during manned entry	Yes	Yes	Yes

(continued)

Table 11-1 Available tools using electromagnetic technology (continued)

	Free-Swimming Deployment		Tethered Deployment		Manual Deployment	
	Neutrally Buoyant Free-swimming	**High-resolution Free-swimming**	**Robotic Crawler**	**Tethered In-line Inspection Tools**	**Handheld Scanner**	**Manned Cart**
Dewatering Required	No	Yes	Not required for crawler	Varies	Yes	Yes
Access	12-in. minimum tap	Through pig launchers or on-size side laterals	16-in. minimum tap	Typically on-size with larger pipe sizes accessible through manholes	Man-size entry point	Pipe diameter dependent. Man-sized entry point required for manned cart
Propulsion Method	Internal water pressure	Internal water pressure	Robotic crawler or manned cart	Outbound tool launch is typically accomplished with air pressure, retrieval by tether.	Manual	Rodding, winching, manual
Other	Requires approx. 1 ft/s flow. 200 psi max	Either wet or dry, tools are pressure sealed from 450 psi to 1,000 psi	For self-propelled tools, 100 psi max. no flows. For manually propelled cart, confined space ready—ventilation, access, and monitoring of personnel	Velocity is controlled by winching speed	Clean and dewatered pipe Confined space readiness—ventilation, access, and monitoring of personnel	Clean and dewatered pipe
Range	Up to 20 h traveling at 80% of flow rate	Up to 15-mi range.	1.5 mi of tether. Up to 3 mi per insertion (u/s and d/s) or walkable distance per day (1–3 mi)	Range limited by wireline length (typically in the 1- to 1.5-mi range).	Pipe diameter–dependent	Pipe diameter–dependent

(continued)

Table 11-1 Available tools using electromagnetic technology (continued)

	Free-Swimming Deployment		Tethered Deployment		Manual Deployment	
	Neutrally Buoyant Free-swimming	High-resolution Free-swimming	Robotic Crawler	Tethered In-line Inspection Tools	Handheld Scanner	Manned Cart
Bend Criteria (D = pipe diameter)	≥24-in. butterfly valves okay. 90-degree bends okay. Travel distance dependent on flow.	Elbows up to 90 degrees conforming to the AWWA C110 (2012) and AWWA C153 (2011) standards with bend radii less than 1 pipe diameter—typically 0.6-0.7D)	1D and 90-degree bends in diameter ≥24 in. Travel distance affected by valves, debris, mud, slopes, bends, and access by personnel.	Elbows up to 90 degrees conforming to the AWWA C110 (2012) and AWWA C153 (2011) standards with bend radii less than 1 pipe diameter—typically 0.6-0.7D)	Any size	6 in. and larger; >16-in. pipes, 90-degree bends okay. When tethered, bends (number, degrees, location to access point) are the primary driver for range variance
Typical Daily Productivity	Up to 20 h traveling at 80% of flow rate.	4.5–5 mi/d	1–3 mi/d	0.5 mi	Tens to hundreds of feet per day	Tens to hundreds of feet per day
Staffing	Varies by field conditions	2-man crew for up to 12 h, 4-man crew for longer runs	Varies by field conditions	Varies by field conditions	Minimum 2-person operation	Minimum 2-person operation
Defect Type	Localized wall loss and corrosion. Groupings of pits	Localized wall loss, general wall loss, graphitization	Localized wall loss and corrosion. Groupings of pits	Localized wall loss, general wall loss, graphitization	Localized wall loss, general wall loss, graphitization, lateral and circumferential fractures	Localized wall loss, general wall loss, graphitization, lateral and circumferential fractures
Defect Size	3 in. × 3 in. × 30% wall loss	Min resolution ⅜ in. × ⅜ in. × 30% wall loss	3 in. × 3 in. × 30% wall loss	Min resolution ⅜ in. × ⅜ in. × 30% wall loss	5–10% wall loss or iron leaching below sensor footprint, minimal discontinuity detection associated with fractures, wall delamination	5–10% wall loss or iron leaching below sensor footprint, minimal discontinuity detection associated with fractures, wall delamination
Defect Location	Longitudinally ±2.5% and ±15 degrees circumferentially	Longitudinally ±1% and 1.8 in. or less circumferentially	Longitudinally ±2.5% and ±15 degrees circumferentially	Longitudinally ±1% and 1.8 in. or less circumferentially	Sensor size–dependent—within ½, 1, or 2 in.	Sensor size–dependent—within ½, 1, or 2 in.

(continued)

Table 11-1 Available tools using electromagnetic technology (continued)

	Free-Swimming Deployment		Tethered Deployment		Manual Deployment	
	Neutrally Buoyant Free-swimming	**High-resolution Free-swimming**	**Robotic Crawler**	**Tethered In-line Inspection Tools**	**Handheld Scanner**	**Manned Cart**
Data Collected	Relative wall thickness. Identification of localized wall loss by volume, area, and percent wall loss	Quantification of wall thickness, location of pipeline features, quantification of bends	Relative wall thickness. Identification of localized wall loss by volume, area and percent wall loss	Quantification of wall thickness, location of pipeline features, quantification of bends	Detecting changes in BEM signal intensity or condition resulting from wall loss graphitization, fractures, or delamination	Detecting changes in BEM signal intensity or condition resulting from wall loss, graphitization, fractures, or delamination
Limitations	Defect detection near joints/features is limited. Wall thickness over ½ in. reduced detection.	Reduced sensitivity at fittings (valves, flanges, etc.)	Defect detection near joints/features is limited. Wall thickness over ½ in. reduced detection.	Reduced sensitivity at fittings (valves, flanges, etc.). Range limited by bends	Inability to detect isolated pits, determination of external versus internal corrosion	Inability to detect isolated pits, determination of external versus internal corrosion
Advantages	Minimal service disruption. Pre-cleaning typically not required. Can pass inline valves. No dewatering required	No dewatering required; will scan long lengths in single deployment; produces high-resolution actionable data needed for prevention of failures	Visual inspection can be carried out. Maximum control over tool. Sonar and laser profiling also available. Manned cart can be rapidly disassembled in pipe to pass valves, tees, or sharp bends at their location.	No dewatering required; will scan long lengths in single deployment; produces high-resolution actionable data needed for prevention of failures.	Scanning through thick coatings or linings, ability to assess condition at time of scanning, ease of operation, accurate flaw positioning, real-time data display	Scanning through thick coatings or linings, ability to scan all pipe diameters and navigate 90-degree bends, ease of operation, accurate flaw positioning, real-time data display
Disadvantages	Flow control required (~1 ft/s). Turbulence may affect data quality. Lower resolution	Flow control and minor cleaning needed. Steep learning curve: in-line inspection procedures and access requirements are common to oil and gas industry but new to most municipal pipeline operators	Shutdown and depressurization at insertion necessary. Travel distance affected by debris, mud, slopes, and bends. Personnel enter under confined-space permitting	Shutdown and depressurization at insertion necessary. Travel distance affected by debris, mud, slopes, and bends. Personnel enter under confined-space permitting	Speed of data acquisition. Pipe shutdown and dewatering is essential when internal inspections are required	Speed of data acquisition, pipe shutdown, and dewatering requirements

BEM: broadband electromagnetic, BWP: bar-wrapped concrete cylinder pipe, CIP: cast-iron pipe , DIP: ductile-iron pipe, EM: electromagnetic

BEM can be used both internally and externally on pipes to detect internal and external flaws and is especially applicable for graphitic corrosion in cast and ductile pipes. It is equally applicable for thin- and thick-walled ferrous pipes and does not require intimate contact with the wall under inspection, making it suitable for surfaces that are lined or coated or exhibit internal scaling or external encrustation.

The basic BEM antenna has both transmitter and receiver coils. Both coils are wound coaxially and, based on the geometry of the pipe, selected antenna can be either a single receiver antenna or multiple receiver antennas to allow for speed of scanning within available space. The assessment is of the wall immediately below the footprint of the antenna.

The transmitter coil is driven with a relatively low-frequency sinusoidal current to produce a magnetic field. This magnetic field induces strong in-wall eddy currents to flow in the area of the antenna footprint. These eddy currents, in turn, produce their own EM field that is detected by receiver coils. By measuring the strength of the voltage induced in the receiver coils, one can identify changes in the wall under inspection. The strength of the EM field created by the wall is generated by the volume and consistency of the wall immediately below the antenna footprint providing for unambiguous locating of flaws or wall reduction.

The BEM antennae are introduced to a pipeline along an exposed surface or via an open end or "launch barrel." Externally, the antenna is placed on the pipe wall manually or with the aid of a housing containing numerous antennae sometimes lowered into an excavation on the end of a mast. For internal scanning, the antenna can also be placed on the pipe wall manually or with the aid of probes that can be pushed by hand or rods or pulled through with the aid of a winch. Internal inspections require the pipe to be off-line, dewatered, and cleaned. Additional information is provided in Table 11-1.

State of the Practice and Application

This section describes in-pipe EM tools that are commercially available and tools that are expected to be released in the near future.

The application of each technology is limited to certain materials, diameter ranges, and linings for which the tool was designed. A summary of each technology is included in Table 11-1. The descriptions of each row in this table are as follows.

- Tool Description
  - *Tool type* – the generic description name of the tool (see column headings).
  - *Pipe wall intimacy* – description of the typical distance between the tool and pipe wall.
- Applicability
  - *Material* – the materials that the tool can assess.
  - *Diameter* – the range of diameters that the tool can assess.
  - *Lining/coating* – the maximum thickness and compatible materials of linings or coatings through which the tool can assess.
- Access and Preparation
  - *Outage required* – "yes" indicates that the pipe needs to be taken out of service to perform the inspection; "no" indicates that the pipe can remain in service during the inspection.
  - *Dewatering required* – "yes" indicates that at least a portion of the pipe diameter needs to remain dry for an inspection.
  - *Access size* – minimum clear diameter required for the tool to be inserted into the pipe.

  - *Propulsion method* – the means by which the tool is moved through the pipe.
  - *Other* – additional pipe preparation items required or access limitations.
- Range and Productivity
  - *Range* – the distance that can typically be inspected by the tool from the insertion point.
  - *Range criteria* – any pipe characteristics that may impact the range.
  - *Typical daily productivity* – the typical total inspection distance capable per day once the tool is in the pipe. The time required to prepare for insertion of the tool into the pipe can vary significantly based on field conditions and has therefore been excluded.
  - *Staffing* – the typical minimum number of recommended onsite personnel to perform the inspection, including owner's staff and tool provider's staff.
- Output Data
  - *Defect type* – a description of the types of defects that can be observed.
  - *Defect size* – a description of the minimum size of defects that can be observed.
  - *Defect location* – a description of the location of the defect(s) that the tool can detect along the pipe longitudinally, circumferentially, or through wall thickness.
  - *Data provided* – describes the format and content of the data provided.
- Advantages and Disadvantages
  - *Limitations* – describes additional limitations of the tool.
  - *Advantages* – describes additional advantages of the tool.
  - *Disadvantages* – describes additional disadvantages of the tool.

DEFECT RESOLUTION CONSIDERATIONS

When selecting a technology for assessing the pipe wall, the resolution of data required for the application should be considered. The EM tools provide different resolutions that will result in varying levels of confidence for decision making purposes. Resolution considerations will depend on the risk of pipeline failure, which is a factor of the consequence and the probability of failure of the pipeline. Higher risk pipelines may justify the need for higher-resolution data while lower risk pipelines may be sufficiently assessed using lower-resolution tools.

EM technology generally provides a relative measurement of the thickness of pipeline wall. However, BEM devices are better at detecting fractures than RFT systems. Resolution is affected by a number of factors. Pipe material and thickness can change the sensitivity of the EM response to wall thickness. Sensor spacing or sizing can also change the resolution.

Table 11-1 describes the types of inspection tools that use EM technology.

ADVANTAGES AND LIMITATIONS

EM technology provides a proven method of accurately and precisely measuring the wall thickness of metallic pipelines. EM tools collect data that can inform short-term repair and replacement decisions as well as long-term capital planning decisions.

The technology is available on a number of platforms that can accommodate a wide variety of operational situations, providing data on the condition of the pipe wall over the full length and circumference of the pipe. The method works with mortar and other common linings and coatings and with scale. However, some scale removal (by pigging) may be needed in order to provide smooth passage of the tool.

Disadvantages include the need for site preparation to provide access to the pipeline for tool insertion or excavations for external access. Some tools require temporary changes in operations or full shutdown of the pipeline.

Every method has limitations, and it is important to understand them when choosing the tool and when evaluating the results of the inspection. Generally, these tools measure the relative wall thickness by evaluating the changes seen in the EM field as it interacts with the pipe wall. The quality of data can be affected by erratic tool movement, excessive tool speed, or interferences from other EM fields like high-voltage power lines. Proper calibration to the specific pipe material and pipe wall thickness is important. Signal distortion can also occur when other ferromagnetic materials are located within very close proximity to the pipe. Joints (bell-spigot) or seams (welded, lock-bar, riveted, etc.) will produce a large response in the data and may mask the presence of wall loss anomalies in the pipe wall near the seam or joint.

REFERENCES

AWWA (American Water Works Association). 2012. C110. Ductile-Iron and Gray-Iron Fittings. Denver: AWWA.

AWWA. 2011. C153. Ductile-Iron Compact Fittings. Denver: AWWA.

Ellison, D.; Spencer, D.; and Brander, R. 2018. Leveraging Data from Non-Destructive Examinations to Help Select Ferrous Water Mains for Renewal. Denver: Water Research Foundation.

Jackson, R.Z.; Pitt, C.; and Skabo, R. 1992. Nondestructive Testing of Water Mains for Physical Integrity. Denver: AwwaRF.

Chapter 12

Magnetic Flux Leakage*

Chris Macey, Chapter Lead, AECOM
Rod Jackson, Co–Chapter Lead, CH2M HILL (now Jacobs)
Margaret Hannaford, San Francisco Public Utilities Commission
Phil Hoyt, Electromechanical Technologies
Allison Stroebele, Pure Technologies

Magnetic flux leakage (MFL) in-line inspection (ILI) provides a detailed full-circumference scan of a pipeline, detecting damage and defects caused by corrosion and other issues. This technology, commonly known as MFL "smart pigging," has been widely used in unlined steel pipelines in the petroleum, petroleum products, natural gas, and liquefied gas transportation industry since the mid-1960s. MFL inspection is the correlation of changes in a magnetic field caused by pipeline defects to the dimensions of the defect. However, the presence of lining in water pipelines has, until recently, limited the accuracy of MFL ILI inspection and hindered development of practical ILI methods for large-diameter, cement mortar lined (CML) steel pipelines.

In 2007, the San Francisco Public Utilities Commission (SFPUC) commissioned the development of MFL pipeline inspection equipment and techniques to inspect its CML steel transmission pipelines for corrosion wall loss. SFPUC was specifically interested in assessing corrosion on 140 miles of large conveyance water lines (56- to 78-in. diameter) that run from the Sierra Foothills through the San Joaquin Valley in California. The result of the efforts was an MFL ILI tool that could be pulled by electric tow vehicles (ETVs) and detect losses in wall thickness starting at about 20 percent with a high degree of accuracy.

MFL is a nondestructive condition assessment that yields high-resolution pipe wall loss data for the full pipeline circumference. The MFL ILI tool has now been used by other

* This chapter is dedicated to Phil Hoyt, a leader in the development of MFL for water pipelines who wrote much of this text. Phil passed away before this manual was completed.

water districts. However, MFL is not applicable to all ferrous pipes. This chapter discusses where this technology is today, current and past data results, and efforts required to perform inspections.

CURRENT STATE OF THE ART

There are only two pipeline inspection technologies that provide direct measurement of pipeline defects. Properly scaled ultrasonic testing (UT) instrumentation measures defect length, depth, and width. Similarly, MFL instrumentation measures defect length and sets close bounds on depth and width. Both can see and quantify small defects. Other technologies use secondary effects to infer the existence of defects and consequently often can only see the effect of larger defects.

For the MLF system to operate, magnets must induce as much magnetism into the pipeline as the pipe wall can hold. A defect will cause the adjacent magnetic field just outside the pipe wall to increase. The increase in magnetic field strength (or magnetic flux density) is detected and interpreted by the MFL tool. Producing the required flux density in the pipe wall when the permanent magnets have been held away from the pipe wall by the lining has precluded the use of MFL inspection in lined pipeline, specifically CML water pipeline. Recent advances in MFL inspection have largely overcome this difficulty, and MFL inspection has been conducted in several CML water pipelines.

Detection capability and defect sizing capability of MFL technology in lined pipeline varies with the characteristics of the particular lining. Available tools are designed for and perform well when the lining is about 0.5 in. thick. Lesser thickness can be inspected. Defects as small as ½ in. in surface dimension with 15 percent wall thickness have been detected by the tool. As a general rule, defects larger than ¾ in. in each surface dimension and depth greater than 20 percent wall thickness are detected and sized, and sizing accuracy increases for larger defects. Pipe with thicker lining and standoff from other causes has been inspected, although that requires modified equipment. Even then, greater standoff lessens field strength, reduces defect detection capability, and decreases sizing accuracy.

The MFL inspection practices discussed in this chapter have been applied to steel pipelines. MFL could theoretically be applied to other ferromagnetic pipe materials such as gray and ductile cast iron, but there are no known commercial applications at this time. Research would be required to assess feasibility and determine instrumentation requirements for inspection.

MFL APPLICATION CONSIDERATIONS AND INSPECTION LOGISTICS

Fundamental MFL Principles

Magnetic flux is simply a synonym for magnetism. An MFL ILI tool uses permanent magnets to induce a magnetic field in the wall of a pipeline. An anomaly such as a corrosion defect disrupts the magnetic field, the ILI tool detects the disruption, and defect analysis correlates the altered magnetic signal to the nature and size of the defect that caused it. The system is simple, direct, and accurate. It is direct inspection and not inferential.

Advanced MFL Principles

MFL uses a circumferential series of dipole magnets to induce a local magnetic field into the pipeline wall as shown in Figure 12-1. In conventional use, flux couplers such as brushes or solid magnet coupling include the pipe wall as part of the magnetic circuit.

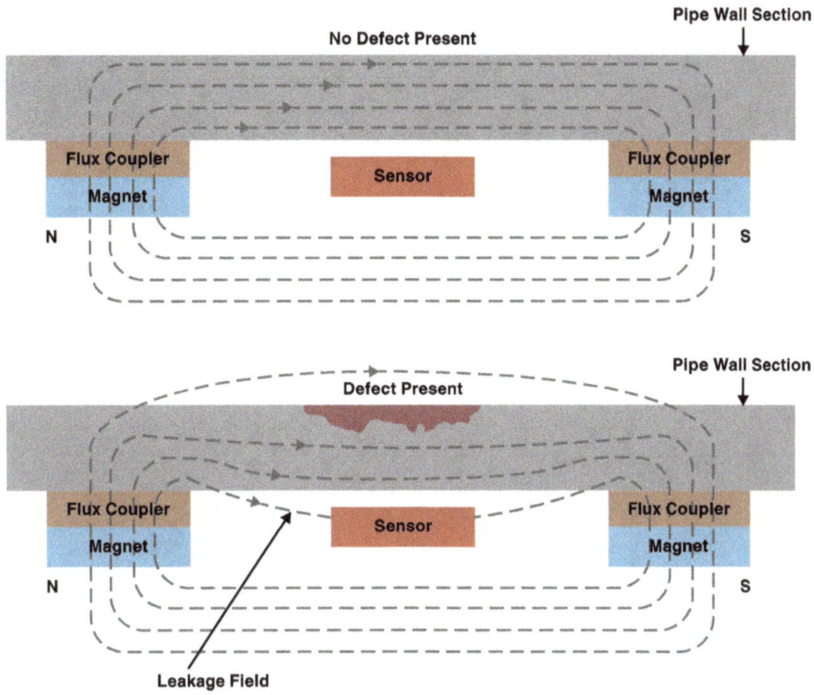

Adapted with permission from Electromechanical Technologies Inc.

Figure 12-1 MFL field in unlined pipe

The primary flux field in the pipe wall is the induced (B) field. When there is a flux field in the wall, there is a parallel lower-density field adjacent to the wall. This lesser field is the leakage field, and it is located where sensors can monitor its strength. At low B field levels, anomalies cause flux levels in the pipe wall to change without significant change in the leakage field. At high flux levels, however, the pipe wall is effectively magnetically saturated, and anomalies cause little change in the B field density. Concurrently, there is significant change in the leakage field, which the sensors see and correlate to properties of the anomaly due to corrosion on either the internal or external wall of the pipeline. For water pipelines, external wall loss due to corrosion is most common.

Figure 12-2 shows the same situation with lining between the magnet and sensor assembly and the pipe wall. In principle, the situations are very similar. The only difference is greater standoff between the magnet/sensor assembly and the pipe wall. Limited coupling through the lining has a severe effect on the magnetic circuit, however, and leakage field flux levels drop to unworkable levels without intervention. There must be intervention within the magnetic circuit and within the sensors to maintain a system in which flux density in the pipe wall is at saturation levels and in which sensors can monitor changes in the leakage field at the pipe wall from a high standoff position.

Figure 12-3 shows a B-H curve typical for pipeline steel. The magnetizing field (H, measured in Oersted) provided by the magnets of Figure 12-1 induces a field (B, measured in Gauss) in the pipe wall being inspected. Although there is not a true asymptote, the B-H curve flattens significantly at its upper limits and the inspected material is said to be in saturation. The flux field is not confined to the pipe wall, and there is a lower level flux field adjacent to the inspected material. This field is the leakage field as shown in Figures 12-1 and 12-2. It is accessible for monitoring. When the inspected material is in the saturation zone, the leakage field changes appreciably if the cross-sectional geometry of the inspected part changes. Corrosion changes the cross-section of pipe wall and changes the leakage

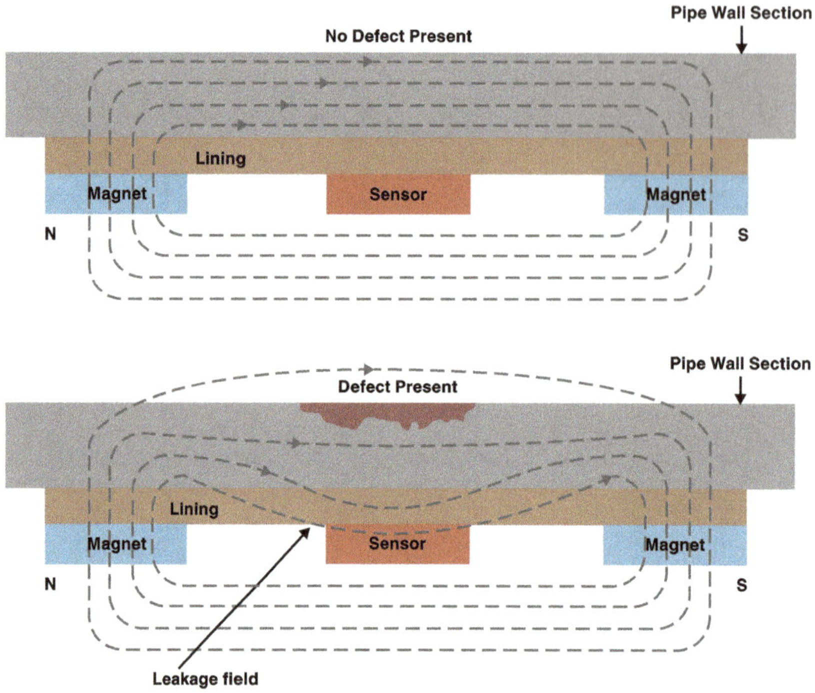

Adapted with permission from Electromechanical Technologies Inc.

Figure 12-2 MFL field in lined pipe

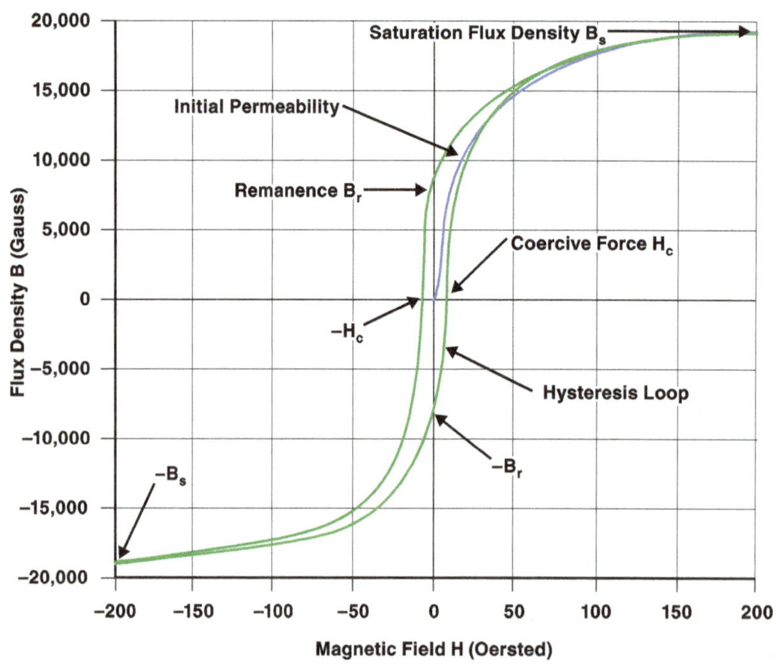

Figure 12-3 B-H curve

field. The process of monitoring leakage field changes in order to characterize changes in the inspected part is MFL inspection.

Magnetic flux density decreases exponentially with distance from the source, roughly following an inverse square law. For that reason, standoff between a magnetic field H and the induced flux density B in pipeline material has a very significant negative effect in lined pipeline. A corollary observation is that the leakage field also decreases in magnitude from its maximum at the pipe wall to points on the inside of the lining where measurements can be taken. Sensors are required that can characterize the leakage field from those measurement points. When techniques are developed to produce the required B field and sensors are developed to monitor the leakage field from accessible locations, MFL is applicable regardless of the standoff distance.

In high-resolution pipeline inspection, the required MFL B field flux density is generally accepted as 17 kilogauss, a point above the significant decrease in slope of the B-H curve.

The ILI Tool Using MFL Technology

A series of magnets are placed uniformly around the inner circumference of the pipe wall to induce a saturation magnetic flux field into the pipe. This is accomplished by matching the tool close to the size of the interior pipe diameter. A corresponding series of sensors also spaced at close intervals (typically about 6 mm) uniformly around the inner circumference of the pipe monitor the leakage field. As the inspection tool moves through the pipe, each sensor traces an individual B-scan (simply stated, a line) along the pipe axis. The data is then a series of B-scans that can be integrated (viewed in the context of one another) to produce a C-scan (i.e., a three-dimensional view) of pipeline defects. Raw data are a series of lines around the circumference running in the direction of the pipe axis. Defects appear as perturbations in these lines, and the combined view of the perturbations represents defects in the pipeline. Analysis provides requisite detail.

A typical MFL tool used in large-diameter, mortar-lined pipeline is shown in Figure 12-4.

Magnetic flux can be thought of as a series of lines; in fact, flux density is often described exactly that way, with a line being a vector representing a given quantity of magnetic flux across a unit of area. Defects that have dimensions normal to the direction of the flux field or these lines are best detected and sized, while defects that are oriented parallel to these lines are more difficult to see. In an axially oriented magnetic field as described above, general corrosion, pitting corrosion, and pinhole defects (definitions for these terms are provided later in this chapter) are readily seen and characterized. Circumferential slotting and grooving present a surface normal to the flux lines and are readily visible to an MFL tool. Circumferential cracks that are open even slightly also present such a surface, but cracks are typically not open, and the flux field will cross a closed or incipient circumferential crack without producing a strong signal. In the axial field described, flux lines tend to "slide by" defects that are strongly axially oriented and axial slotting, grooving, and cracks are less visible.

One MFL technique that has been used in unlined pipeline inspection is to turn the magnetic field 90 degrees and to run the flux field circumferentially. Defects that are oriented along the pipe axis (axial slotting, axial grooving, and open axial cracks) then become the defects that are more readily seen.

Lining tends to obscure defects that are already hard to see in conventional MFL inspection, again such as axial slotting or grooving. However, when there is amplifying intervention in the magnetic circuit and special sensors are provided as discussed previously, MFL becomes not only a viable technology but a superior technology for the inspection of lined pipeline.

Other defects can be common in lined pipeline. Mortar disbondment (perhaps referred to as "delamination" in other sources) can cause a drop in the strength of the flux

Source: San Diego County Water Authority and San Francisco Public Utilities Commission

Figure 12-4 Typical MFL ILI tool in CML pipeline

leakage field that can be misinterpreted as pipe wall damage. However, it has a distinctive signal that is recognizable by a qualified analyst. Lamination of the steel plate comprising the pipe itself has been known to be present. It presents a thin edge to the magnetic flux front and is significantly less visible than other defects. Lamination is usually a manufacturing defect and is relatively infrequent. MFL is not a good tool to find it. When mortar lining also intervenes, detection of metal lamination is difficult to see. There are limited known, practical techniques to find metal lamination in mortar-lined pipelines.

Defects as small as ½ in. in surface dimension with 15 percent wall thickness loss have been seen but are difficult to recognize and quantify through mortar lining. As a general rule, defects larger than ¾ in. in each surface dimension and depth greater than 20 percent wall thickness are detected and sized, and sizing accuracy increases for larger defects. Other mortar lining defects such as mortar drops lost during repair can be mistaken for metal corrosion but have unique characteristics such as signal sign and spread that also help an experienced analyst identify them.

To help visualize MFL data, Figure 12-5 shows a section of MFL B-scans, each B-scan or line produced by a magnetic sensor in a magnetic field as the inspection tool passes along the pipe axis. Sensors are spaced about ¼ in. apart around the circumference of the pipe. Perturbations in the magnetic leakage field caused by defects in the pipeline are seen as vertical excursions from the horizontal norm. These results are from a controlled test in pipe with ½-in. thick wall through ½-in. thick lining. All defects except the hole are 15 percent deep through the pipe wall, and all have round bottoms.

Inherent clarity of data is seen in these 15 percent defects as small as ½-in. diameter. Defects with ¼-in. diameter are visible but are not very pronounced. These same characteristics occur in functioning pipelines, but mortar lining typically has more variation than

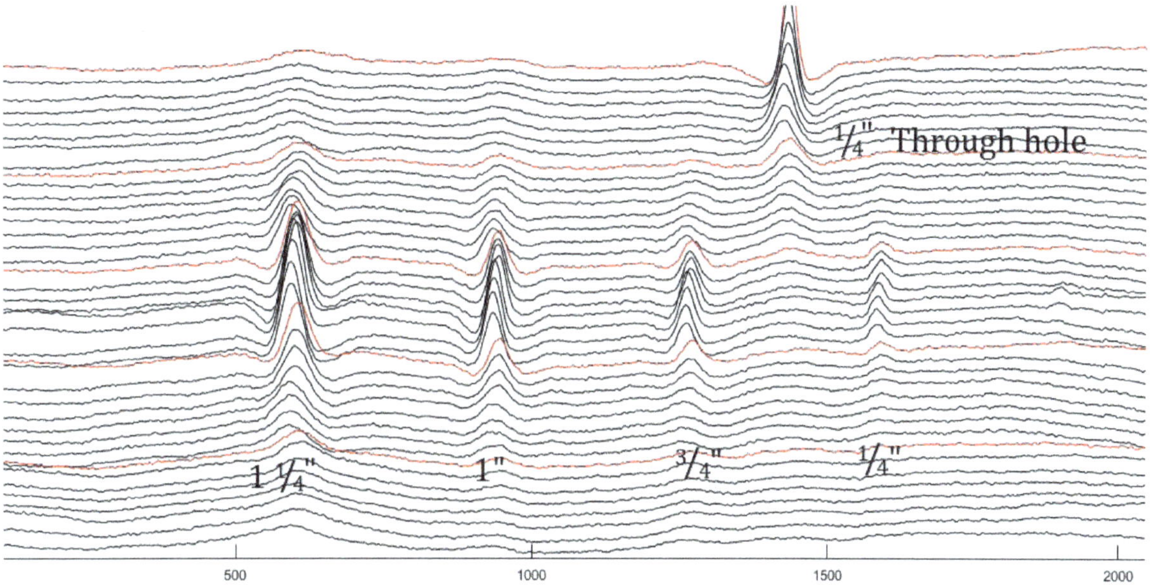

Source: Electromechanical Technologies Inc.

Figure 12-5 Test data in ½-in. pipe with ½-in. mortar lining

this controlled thickness lining. Defects as small as ¼ in. are not very distinct in operating pipelines unless they are deeper, but the hole illustrates that even deep ¼-in. defects are visible.

Inspection Tool Propulsion

Many oil and natural gas pipelines are fabricated to include launchers and receivers that allow inspection tools to be inserted into operating pipelines without disrupting their operation. Inspection tools are then propelled by the product flowing in the pipeline. Typical water pipelines do not have launchers or receivers, although other propulsion methods can be used. One factor driving that situation is that water transmission pipelines are often large, and launchers would be prohibitive from both technical and cost perspectives. Most MFL inspection of water pipeline has been conducted in dewatered pipelines. Access has been obtained through manways through which inspection tools and tow vehicle are inserted. An early challenge in the manufacture of MFL aqueduct inspection equipment was making it flexible enough so that it could be inserted that way. The problem was solved, and both inspection and tow equipment can now be introduced into the pipeline through manways as small as 14 in. × 16 in. Tow equipment cannot produce exhaust into the pipeline, of course, so electric tow vehicles are used.

There are temporary launchers and receivers in use in the oil and gas industry. This equipment can be placed for the inspection and removed when the inspection is complete. That can certainly be done for MFL inspections of water pipelines and has been done for the launching and receiving of electromagnetic tools. Cost and accessibility are both issues.

Examples of MFL ILI Inspection

Figure 12-6 shows results from a 56-in. diameter operating pipeline with two areas of small localized corrosion. Defects are shown in the MFL data and in photographs of the excavated area. Defects in the pair at the left are each about ½ in. in diameter and are about

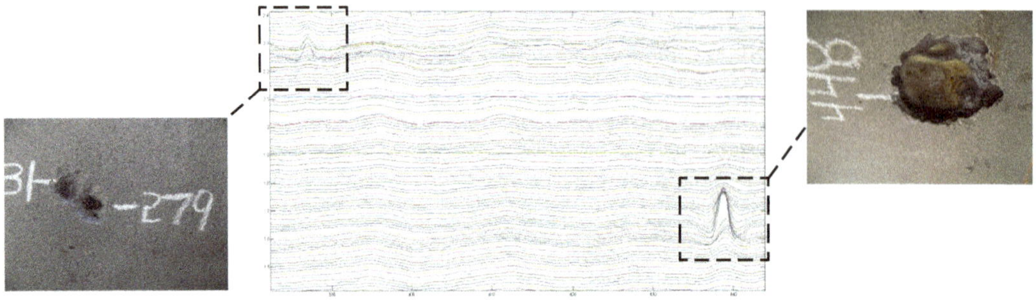

Source: San Francisco Public Utilities Commission

Figure 12-6 Isolated pipeline defects

Source: San Francisco Public Utilities Commission

Figure 12-7 MFL data from severely corroded pipeline and removed pipe showing corroded area

56 percent of the pipe wall deep. Note that the two together produced a single MFL signal. The single defect at the right is about 2 in. in extent, but it is about 90 percent of the pipe wall deep. This particular pipeline had multiple defects of this nature, with this one being the deepest. These defects were on the exterior of the bottom of the pipe.

Correlation between MFL data and actual defects in the wall of a mortar-lined pipe can be excellent. Figure 12-7 shows MFL data from a corroded area in a 56-in. diameter operating pipeline and the actual observed condition of the pipe after removal. The axis of the pipe is horizontal, and the circumference of the pipe is vertical in the images. Interaction is pronounced at the top and along the right side of the MFL image.

These MFL inspection results were used to perform a "desktop" analysis of pipe hoop strength. Through a combination of this desktop analysis and further UT inspection, this section of pipeline was determined to be in critical condition and was replaced with new pipe. The old pipe was recovered and examined. The MFL data compared well with observed condition shown on the right side of the figure. This corrosion was on the bottom exterior of the pipe. Defect depth ranged from nominal to more than 70 percent.

Figure 12-8 depicts corrosion shown in MFL data from a 72-in. diameter pipeline constructed in 1959. To the right side of the figure is the exposed section of damaged pipe found from the MFL data. The corrosion is interior and circumferential and is visible in the 3-ft × 3-ft section of removed mortar lining. Handheld UT found that the upper area

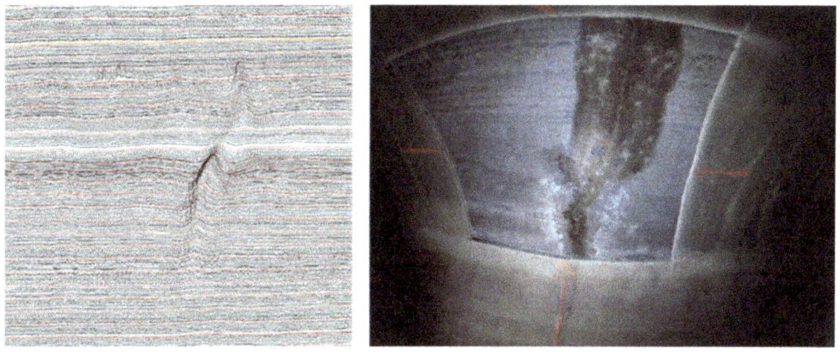

Source: Pure Technologies

Figure 12-8 MFL view of circumferential corrosion and exposed pipe showing circumferential corrosion

of corrosion staining was primarily light surface corrosion without significant wall loss. The lower area is seen to correlate with the MFL data. There is a barely visible crack in the mortar lining below the removed mortar. There was no clear indication of corrosion from visual inspection before the MFL inspection.

DATA OUTPUT AND POST PROCESSING

Definition, Clarity, and Accuracy of MFL Inspection

MFL inspection measures perturbations (e.g., disturbance and irregularities) in the magnetic leakage field. Analysis is required to determine defect dimensions consistent with the magnetic signal. The data are comprised of digital values of the magnetic flux leakage field. A significant advantage is that a three-dimensional (3D) plot of the perturbations in this field looks very much like the pipe wall defect itself. In other words, MFL ILI inspection produces both an image of the defect and detailed dimensions of the defect through post-processing analysis. Both are commonly reported to the owner. Figure 12-9 shows test defects that one water district used in a blind CML test of an MFL tool. This part of the test was a set of five 1-in. and ½-in. defects, all formed with a round bottom at 20 percent wall thickness depth. The images are C-scan plan and 3D views of the same data, rotated 90 degrees from each other. The reader is encouraged to answer the question of which defects are 1 in. and which are ½ in.

Defect Characterization Definition

The Pipeline Operators Forum is a nonprofit association registered in the Netherlands. It comprises 23 major energy and chemical companies organized to promote global pipeline integrity management. One of its major goals is to formulate technical guidelines and specifications for the safe and reliable operation of oil, gas, and chemical pipelines. Those guidelines and specifications have become the backbone requirements for energy pipeline integrity throughout the world, including the United States.

The Pipeline Operators Forum (2016) has established definitions of categories of pipeline defects that have become standard definitions worldwide. They provide meaning that is the same to everyone in the pipeline business regardless of the material transported. To provide standard meaning to pipeline defect definitions, they are used here and are presented in Table 12-1.

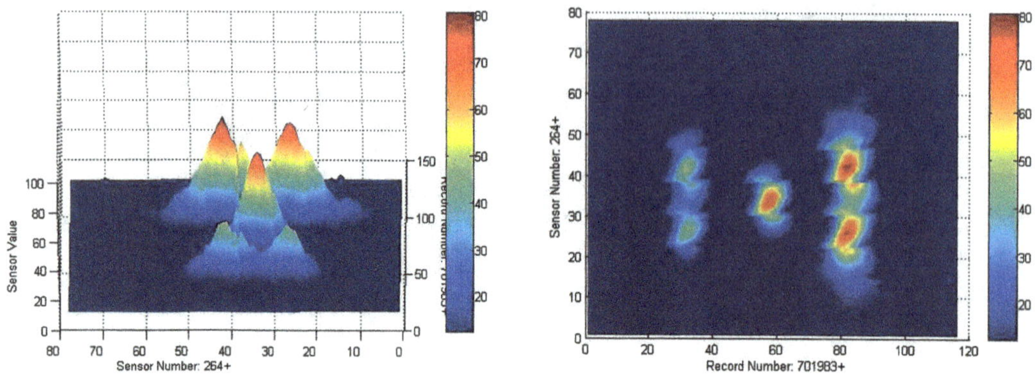

Source: San Francisco Public Utilities Commission

Figure 12-9 Color contour C-scan of blind test defects

Table 12-1 Definitions of pipeline anomalies

Anomaly Dimension Class	Definition	Reference Point/Size for the POD in terms of L x W
General	{[W ≥ 3A] and [L ≥ 3A]}	4A × 4A
Pitting	{([1A ≤ W < 6A] and [1A ≤ L < 6A] and [0.5 < L/W < 2]) and not ([W ≥ 3A] and [L ≥ 3A])}	2A × 2A
Axial grooving	{[1A ≤ W < 3A] and [L/W ≥ 2]}	4A × 2A
Circumferential grooving	{[L/W ≤ 0.5] and [1A ≤ L < 3A]}	2A × 4A
Pinhole	{[0 < W < 1A] and [0 < L < 1A}	½A x ½A
Axial slotting	{[0 < W < 1A] and [L ≥ 1A]}	2A × ½A
Circumferential slotting	{[W ≥ 1A] and [0 < L < 1A]}	½A × 2A

Source: Pipeline Operators Forum

A: a geometrical parameter used to specify the dimension class of metal loss anomalies detected by in-line inspection of a pipeline, L: length of the defect in selected units, POD: probability of detection, W: width of the defect in selected units

The probability of detection (POD) of an anomaly is normally 90 percent for the size references shown, but that standard has not been established for water pipeline. The reason is the presumed reduction in sensitivity occasioned by lining in water pipelines. Nonetheless, the general standard is presented here. Variation from the standard should be available from the inspection provider.

The typical detectible depth of a high-resolution MFL tool for general corrosion is 10 percent of the wall thickness (t), or 0.10t. The typical detectible depth for pitting is 15 percent of the wall thickness, or 0.15t. Both should have a POD of 90 percent. Depth, length, and width sizing accuracy for high-resolution MFL tools are depth ±10–15 percent and length and width accuracy of ±10–20 mm (0.4–0.8 in.). The typical certainty level for sizing anomaly depth, length, and width is 80 percent. Again, variations caused by pipe lining should be available from the inspection vendor.

The Pipeline Operators Forum advises location accuracy of an anomaly for an MFL tool to be 0.15 meters (0.5 ft) from the closest weld. It recommends circumferential location accuracy of the anomaly to be within ±10 degrees of the anomaly. Typical location accuracy for existing tools is ±1 percent of the distance from a reference point and ±3 in. (7.62 cm)

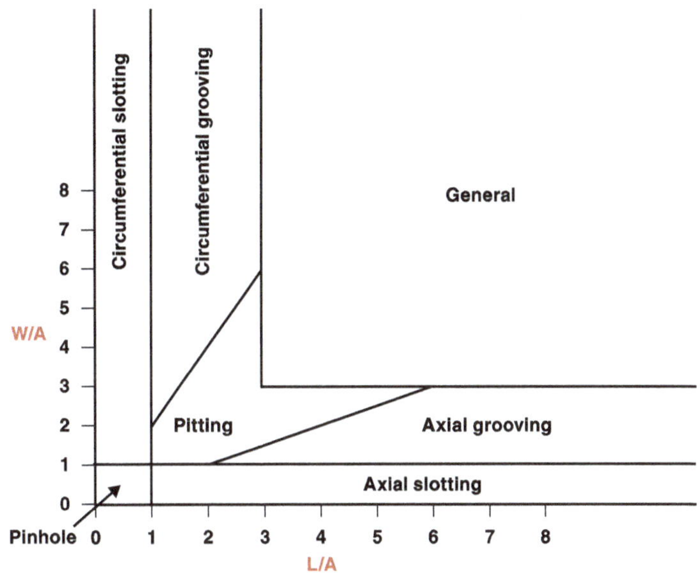

Adapted with permission from Pipeline Operators Forum
A: a geometrical parameter used to specify the dimension class of metal loss anomalies detected by in-line inspection of a pipeline, L: length of the defect in selected units, W: width of the defect in selected units

Figure 12-10 Graphical depiction of pipeline defect anomaly definitions

from the nearest pipe joint. Pipe joints are visible in MFL data and serve as fixed reference points for longitudinal location accuracy. Location around the circumference of the pipe is ±10 degrees. Some MFL tools carry a six-axis, strap-down inertial navigation system that is also used as an inclinometer; some may also have the capability of a full inertial navigation unit with refined accuracy for pipeline location.

Shown graphically in Figure 12-10, these metal loss anomaly classes are defined to specify the metal loss measurement capability of an inspection tool. They are presented as a plot of length along the axis of the pipe versus width around the circumference of the pipe. The parameter A simply makes the plot nondimensional so that it applies to pipe of all wall thickness. If wall thickness (t) is less than 10 mm (0.4 in.), A is defined as 10 mm. If it is greater than 10 mm, A is equal to t. Use of these detailed definitions has further advantages in characterizing defect growth using MFL tools that can provide the detail required to monitor growth.

Performance Specifications

The average specified performance of several conventional MFL ILI tools is given in Table 12-2. Inspection through pipe lining would naturally be expected to reduce these specifications. At the present time, there is not sufficient data available to provide accurate statistical analysis for specifying these values in lined pipeline, but there is a general feel for what they might be based on the experience of personnel who have worked with both unlined and lined pipeline. Within necessarily wide variation, these estimated values are also given in Table 12-2.

Table 12-2 Estimated performance specifications

Performance Specifications	Average in Unlined Pipe	Estimated in CML Pipe
GENERAL		
Depth of detection at POD = 90 percent	0.10t	0.15t
Depth sizing accuracy at 80 percent certainty	±0.10t	±0.15t
Length sizing accuracy at 80 percent certainty	±13 mm	±15 mm
Width sizing accuracy at 80 percent certainty	±15 mm	±25 mm
PITTING		
Depth of detection at POD = 90 percent	0.10t	0.15t
Depth sizing accuracy at 80 percent certainty	±0.10t	±0.15t
Length sizing accuracy at 80 percent certainty	±13 mm	±13 mm
Width sizing accuracy at 80 percent certainty	±13 mm	±20 mm
AXIAL GROOVING		
Depth of detection at POD = 90 percent	0.11t	0.20t
Depth sizing accuracy at 80 percent certainty	±0.14t	±0.20t
Length sizing accuracy at 80 percent certainty	±13 mm	±25 mm
Width sizing accuracy at 80 percent certainty	±16 mm	±25 mm
CIRCUMFERENTIAL GROOVING		
Depth of detection at POD = 90 percent	0.10t	0.15t
Depth sizing accuracy at 80 percent certainty	±0.10t	±0.15t
Length sizing accuracy at 80 percent certainty	±13 mm	±15 mm
Width sizing accuracy at 80 percent certainty	±13 mm	±25 mm
OTHER		
Operating temperature range	0–60°C	0–65°C
Maximum operating pressure	2,185 psi	2,000 psi
Tool speed	3.5 m/s	1–1.5 m/s
Maximum operating time	180 h	15 h/day
Maximum inspection length	440 km	5 km/day
Sampling frequency speed	1,075 per second	1,400 per second
Minimum bore	15 percent reduction	±12 percent to ±8 percent from nominal diameter

Source: Electromechanical Technologies Inc.
CML: cement mortar lined, POD: probability of detection, t: thickness

PLANNING AND DELIVERING A QUALITY MFL PROJECT

At the present time, ILI by MFL technology is considered most applicable to water transmission pipelines constructed from steel pipe. Joint type is not a major factor in the feasibility of inspection; although most pipelines that have been examined using MFL have had welded or riveted joints, gasketed-style joints would not preclude the use of MFL. The nature of pipe lining is a consideration for MFL, particularly for pipelines with CML where the magnetic circuit and sensor system in the tools must be suitable for the thickness of the lining.

There are other detailed considerations for planning a pipeline condition assessment by MFL ILI technology. Key considerations are as follows.

- **Project team:** MFL inspection will usually require several team members with a variety of functions and skills. The services provided by an MFL inspection company may be limited to supplying the tools, resultant data, and report of findings. Other engineers and technicians will often plan the project, develop engineering controls to be used during the course of the work, and interpret or apply the results. A construction services contractor will usually be retained to open and close the pipeline, dewater low-lying areas, assemble and disassemble the tool, and propel the tool through the pipeline. Assignments of a project manager and field superintendent are advisable to coordinate activities and maintain schedule, particularly if the inspection is complex or occurs over a prolonged period of time.
- **Service outage:** The pipeline must be taken out of service for the time required to complete the inspection and related activities. It should be recognized that even drained pipelines may contain water at low points along the line, and water resistance is a consideration, especially for the tool electronics. The current tool can handle no more than a few inches of water at the bottom of the pipe. Disinfection of pipelines may be required before return to service.
- **Pipeline access:** The pipeline must be accessible at suitable points for insertion and removal of tools, such as at access manways, valve stations, or even removed sections of pipeline. A straight pipe section of suitable length will generally be required for insertion of the tool. Tools for large-diameter pipelines will likely require assembly and disassembly in the pipeline, and provisions for worker safety are mandatory. Safety provisions include ventilation monitoring, confined space permits, traffic control, personal protective gear, and egress, even in the presence of the inspection tool.
- **Pipeline geometry:** The pipeline that will be inspected must have internal diameter, directional changes, and circularity that are within the tolerance of the MFL ILI tools. This is necessary to minimize the possibility that tools will become stuck in the pipe or that the pipe will be damaged. A preliminary inspection is recommended to confirm pipeline internal diameter, bends, and circularity before attempting to move the MFL tool through the pipeline. Geometry tools may be included in the services provided by the MFL inspection company.
- **Contingency planning:** It should be recognized that unanticipated problems may be encountered during the inspection, such as malfunction of the tools or transport system or the tool becoming stuck in the pipe. Possible problems that may be encountered should be identified and contingency plans made to address them before inspection.
- **Data handling:** MFL inspection will produce large amounts of data that must be transferred from the tools, stored, analyzed or interpreted, and retained. Arrangements must be in place to ensure that the data are preserved throughout the process. In addition, there is a need to promptly screen the inspection data for quality assurance and to identify locations where significant damage is indicated and there is opportunity to make repairs while the pipeline is out of service. The MFL inspection company will typically provide a final report with a table of locations on the pipeline where metal loss is indicated, along with detected features such as joints and old repairs.
- **Validation:** It is advisable to validate, or verify, the results of the MFL inspection to confirm detected indications of damage due to corrosion or impact to the pipeline. This step is helpful in proving the accuracy of the inspection, especially if there are indications of significant damage that require prompt action to prevent leaks

or pipeline failure. Validation may include excavation and examination of the pipeline or ultrasonic thickness mapping of the pipe wall from inside the pipeline to avoid excavation. Since most defects in lined pipe are external, verification from the interior usually takes place through the pipe wall. The operator can strip the lining, find and size the defect from the pipeline interior without excavation, and verify the MFL call using ultrasonic evaluation. Removing the lining and using handheld ultrasonic instrumentation is possible, but small defects seen with MFL may be hard to find that way. With the lining removed, scanning a larger area at closer intervals with instrumentation such as automated ultrasonic testing (AUT) produces better results. A better verification technology once the lining is stripped may involve an oblique wave front ultrasonic test such as phased array ultrasonic evaluation.

- **Budgeting for MFL Inspection:** The direct cost of MFL water pipeline inspection is usually competitive with other methods of water pipeline ILI technology, including limited excavation-based approaches such as NACE International SP0502, "Pipeline External Corrosion Direct Assessment Methodology." The cost of MFL inspection for a particular project, like all ILI technologies, is greatly affected by numerous factors including the number of mobilizations required to insert and remove inspection tools, the length of the inspections, the extent of labor and equipment required to access the pipeline, the diameter of the pipeline, and other system modifications if necessary to accommodate the inspection. Based on projects completed to date, the cost for MFL inspection of large-diameter (54- to 78-in.) pipelines can be expected to range from \$20 to \$30 per linear foot (exclusive of third-party and other costs).

Integrity Management

A single MFL inspection that is properly conducted will characterize damage that exists in the pipeline at the time of the inspection. Repeated MFL inspections over a period of time can be used to determine the rate of corrosion of the pipeline and monitor the pipeline for external damage such as may occur from other construction or farming operations in agricultural areas. Repeated inspections will also monitor the effectiveness of corrosion protection such as cathodic protection and show areas where protection may have been compromised.

REFERENCE

Pipeline Operators Forum. 2016. Specifications and Requirements for In-line Inspection of Pipelines. https://www.pipelineoperators.org/downloads-links/ (accessed Jan. 9, 2019).

Chapter 13

Condition Assessment of PCCP

Bethany McDonald, Chapter Lead, Black & Veatch
Mike McReynolds, Brown and Caldwell
Rasko P. Ojdrovic, Simpson, Gumpertz, and Hager
Ashan McNealy, Pure Technologies
Shaoqing Ge, American Water

Prestressed concrete cylinder pipe (PCCP) has been problematic in some cases because it can fail without warning and in a catastrophic manner. As a result, much attention has been paid to the assessment of PCCP, and many tools and techniques specific to this type of pipe have been developed. Some of these tools are also applicable to other types of materials (particularly bar-wrapped steel cylinder type pipe).

This chapter provides an overview of how to assess the condition of PCCP and some strategies for managing PCCP. These topics are covered in additional detail in the references at the end of the chapter.

PCCP CONSTRUCTION

PCCP consists of a concrete core, steel cylinder, prestressing wires, and mortar coating. PCCP has two general types of construction (Figure 13-1):

- Lined cylinder pipe (LCP) has a steel cylinder lined with a concrete core. This pipe type is typically used for 16-in. to 48-in. diameter applications.
- Embedded cylinder pipe (ECP) has a steel cylinder embedded in a concrete core. This pipe type is typically used for 48-in. to 144-in. and larger-diameter applications.

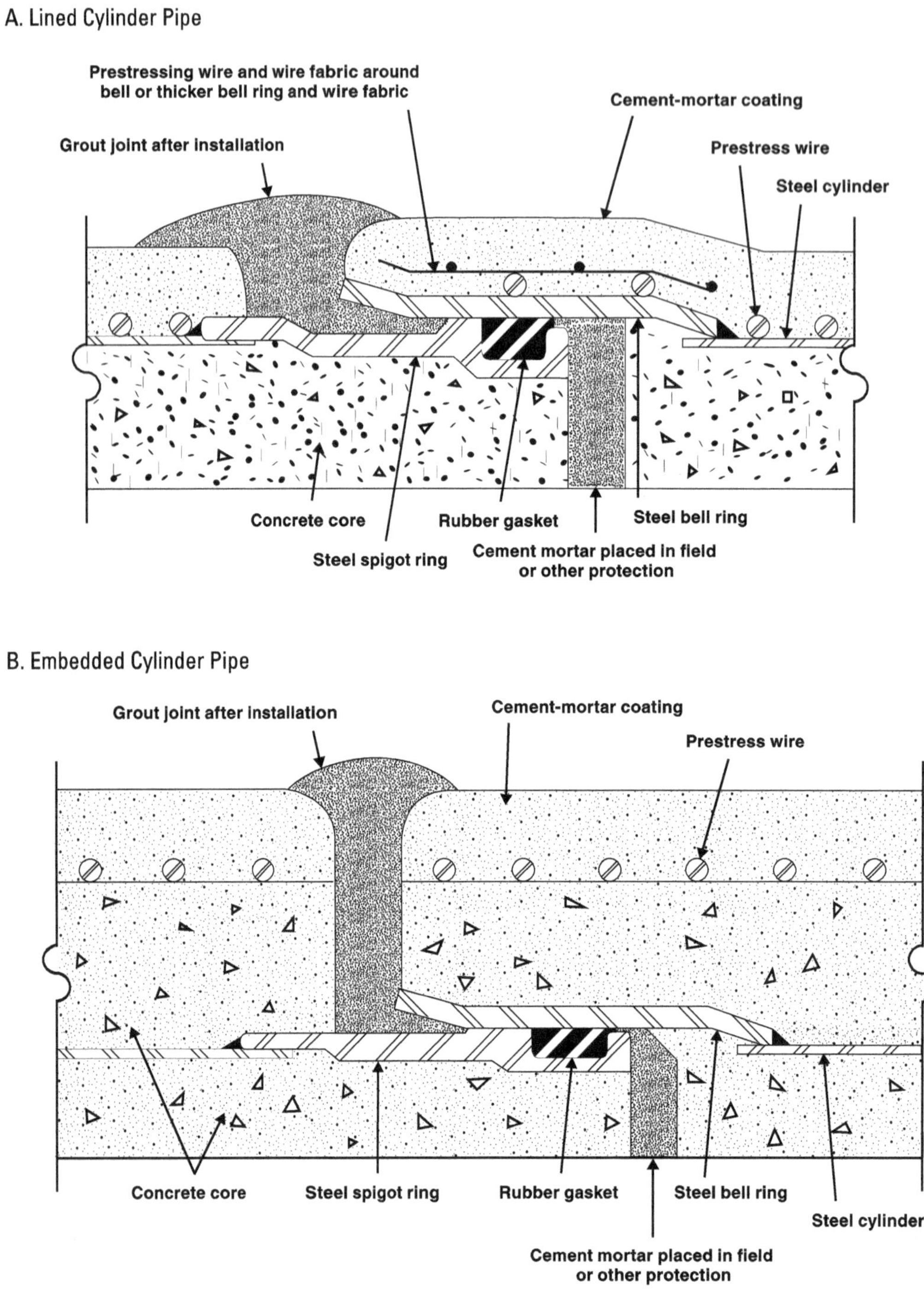

Source: AWWA M9 (2008)

Figure 13-1 Pre-stressed concrete cylinder pipe

In ECP-type PCCP, high-strength wires are wrapped helically around the concrete core. In LCP-type PCCP, the wires are wrapped directly onto the steel cylinder. A cement-rich mortar coating is applied to protect the wires from corrosion. The steel cylinder is the pipe's water-impervious membrane and provides some structural strength to the pipeline, particularly for pipelines designed for relatively low operating pressure. However, for moderate- to high-pressure PCCP, the wires are the primary structural component.

The concrete core thickness is at least one-sixteenth the inside pipe diameter. The pipe is considered to be rigid as it undergoes minimal deflections under earth load. The pipe is typically manufactured with steel bell and spigot joints and rubber gaskets. Electrical continuity is inherent within LCP-type PCCP because the wires are wrapped directly on the steel cylinder. However, in ECP-type PCCP, shorting straps can be present between the concrete core and wires to provide electrical continuity between the wires. For both types of PCCP, electrical continuity between pipes is provided either by welding joints or by constructing joint bonds at each joint.

PCCP is designed for combined loads including pipe and fluid weight, working and transient pressure, earth load, and live load. PCCP is designed in accordance with AWWA C304, "Design of Prestressed Concrete Cylinder Pipe" (before 1992, PCCP was designed in accordance with AWWA C301, "Prestressed Concrete Pressure Pipe, Steel Cylinder Type"). PCCP is also designed for longitudinal forces due to hydraulic pressure-induced thrust at elbows, tees, bulkheads, and other fittings and Poisson's effects of internal pressure and temperature in accordance with AWWA M9, *Concrete Pressure Pipe*. If necessary for complex bends, a finite element software program is used to design the pipe, taking into consideration pipe–soil interaction.

PCCP DETERIORATION AND FAILURE MODES

Degradation and failure of PCCP may occur in a circumferential or longitudinal mode. Degradation in the circumferential mode is typically in the form of corrosion or hydrogen embrittlement of the wires. The failure process includes initiation of distress followed by cracking and delamination of protective mortar coating, leading to exposure of embedded steel wires or cylinder to the environment. Damage to the coating can occur during construction, be induced by a third party, be caused by poor bedding, or be caused by differential settlement between the pipe and encasement or structures. The quality and condition of the mortar coating over the wire are very important to protect the wires from an aggressive environment. Additional corrosion can cause wire breakage, cracking of the concrete core, corrosion of the steel cylinder, and eventual failure under working or transient pressures. Failure may or may not be preceded by leakage depending on pipe design and pressures.

Another dominant failure mode is brittle breakage of high strength wires (Classes III and IV) due to corrosion-induced hydrogen embrittlement. This type of failure can be caused by stray electrical currents from nearby utilities that are under cathodic protection, cathodic protection of the PCCP itself, or poor manufacturing of the prestressing wire. Sensitivity to embrittlement occurs more frequently in higher-strength prestressing wires and is affected by the wire manufacturing process, including drawing temperature, speed, and reduction of area. Class IV wire is especially vulnerable to this type of corrosion. Class IV prestressing wires were manufactured between the early 1970s and early 1980s by one pipe manufacturer using a process that resulted in higher-strength wire that still adhered to the standards. Class IV wire manufacturing was ultimately proven through analysis of subsequent PCCP failures to result in dynamically strain-aged wire susceptible to longitudinal splitting and brittle breakage. The history of wire manufacturing, testing, and problems has been thoroughly reported (Zarghamee et al. 2012, Romer, et al. 2008). Testing has established that all Class IV wires are and some Class III wires may be susceptible to hydrogen embrittlement. Class I and II wires are typically not susceptible to embrittlement, except resulting from corrosion.

In the longitudinal mode, concrete pressure pipe may fail as a result of pipe movement caused by differential settlement, inadequate hydraulic thrust restraint, Poisson's effect of pressure, thermal loads, nearby blast or vibration loads, or seismic loads.

The failure process may involve opening of joints or cracking of the concrete core or tearing of the steel cylinder with or without corrosion, and failure with or without prior leakage.

Some of the more typical PCCP damage mechanisms include the following.

- Corrosion or hydrogen embrittlement of wires, allowing wire breakage and loss of prestress. The loss of prestress may lead to longitudinal and circumferential cracking in the concrete core and debonding between the inner concrete core and steel cylinder.
- Pipe overload may result in longitudinal cracking near the invert or crown or coating cracking at the springline. The cracked coating makes the wire susceptible to corrosion and breakage, resulting in loss of prestress.
- Longitudinal and circumferential cracking of the core and loss of prestress may allow water pressure to get behind the inner concrete core/liner and push the steel cylinder out, resulting in separation at the inner concrete core/liner to steel cylinder interface.
- Inadequate thrust restraint may result in failure at the joint due to pipe movement, tearing of the cylinder due to yielding, tearing of the weld at the bell or spigot, and circumferential cracking of the core.
- Improper bedding can cause cracking of inner concrete core/liner and buckling of the steel cylinder, both of which are visually detectable, or cracking of the mortar coating, which leads to corrosion of the wire and loss of prestress.

INSPECTION AND ASSESSMENT

Desktop Assessment

The first step in assessing PCCP is often a risk assessment, taking into consideration: (1) the manufacturing era, (2) the manufacturer, (3) the wire type, (4) the pipe's performance history, (5) the corrosivity and stability of the soils, (6) operating pressures (including surges) and (7) the potential consequences of failure.

As shown in Figure 13-2, pipes manufactured between approximately 1969 and 1980 have been particularly problematic due to the use of Class IV wire (as described previously) and a general lessening of the manufacturing standards. When failures began to be a large issue, the standards were subsequently strengthened.

Direct Assessments

Direct assessments are defined as those that directly assess the condition of the steel, concrete, or mortar constituents of PCCP. The most effective direct assessments seek to identify indications of PCCP damage mechanisms.

Visual and sounding inspection. Visual and sounding inspection was one of the earliest methods of assessing PCCP condition and remains a widely accepted method of PCCP assessment. Owners began using visual and sounding inspection to identify pipe with severe loss of prestress and other anomalies in the 1980s. Many owners continue to use both visual and sounding inspection and electromagnetic (EM) inspection to assess PCCP condition.

Internal inspection consists of visual inspection of pipe interior for cracking, condition of joints and other anomalies, and sounding of the concrete core using a 1- to 2-lb hammer, metal rod, or another suitable striking device. Internal inspection methods and their limitations have been thoroughly described in the literature (Zarghamee et al. 2012). To conduct this type of inspection, pipelines are entered by a two- or three-man crew. The inspection crew is supported by a top-side confined-space entry support crew.

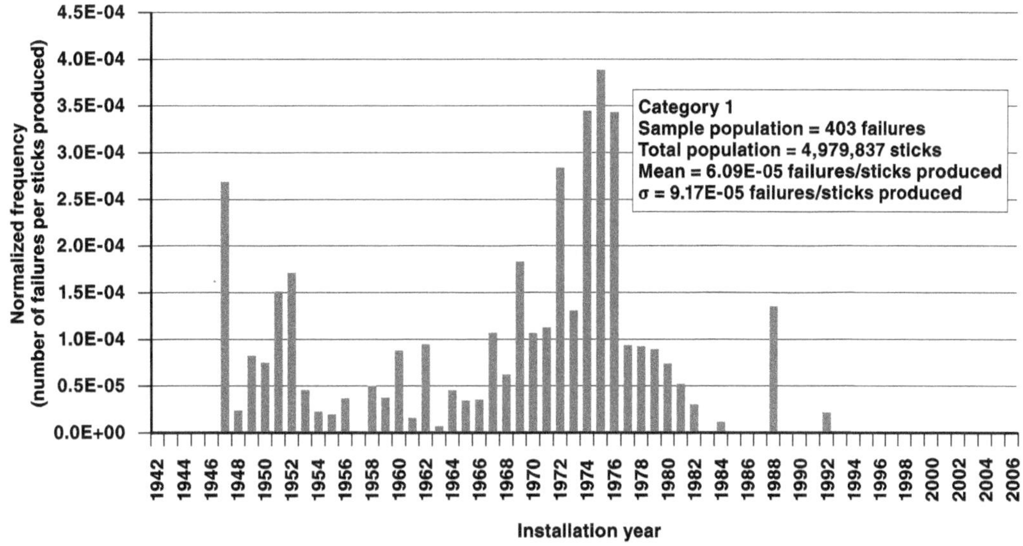

Adapted with permission from Romer et al. (2008)

Figure 13-2 Break rates for PCCP manufactured in different eras

The inspection crew walks (or crawls) through the pipe looking for evidence of concrete damage. Crew members look for visible cracking. The inspection crew sounds the pipe by tapping it using a metal rod, hammer, or other striking device to detect the extents of hollow-sounding areas. The inspection crew records the location and type of damage found.

Figure 13-3 shows typical tools and equipment used in performing visual and sounding inspections. In visual inspection of smaller-diameter PCCP, two people are typically sufficient, as the space is limited. A photographic record of significant anomalies observed should be documented in the inspection report. Opinions differ regarding whether the entire pipe circumference or just the pipe springline and potentially concerning areas must be sounded. Ideally, with sufficient inspection time and budget, the entire circumference should be sounded. However, in many cases a limited pipeline shutdown is available and inspection resources are limited. In these cases, it is generally agreed that sounding of the pipe springline and near areas of concern is sufficient. Refer to Chapters 8 and 9 for general and additional information on internal visual inspection.

Internal inspection can identify pipes in a very advanced level of deterioration as defined by broken wires. Longitudinal cracking and hollow-sounding areas can be caused by separation at the inner concrete core/liner to steel cylinder interface.

Figure 13-4 shows an example of longitudinal cracking surrounded by a hollow-sounding area. High correlation exists between severe distress and simultaneous occurrence of hollow sounds and longitudinal cracking, but not with individual occurrence of hollow sound or longitudinal cracking. Cracking or hollow sounds may occur in otherwise good pipe due to surge, differential shrinkage, local dents in the cylinder, wrapper plates associated with outlets, and other causes. Since detecting the hollow sound of such pipe is subjective, the inspectors typically err on the safe side. The likelihood of identifying anomalies with no relation to pipe deterioration is significant. Because identification of hollows and other signs of pipe deterioration can be subjective, inspector experience is very important in obtaining reliable results. Visual inspection and sounding typically fail to identify pipes that have just a few broken wires, where the damage is insufficient to cause concrete core cracking and separation from the steel cylinder. Where possible, external inspection and verification of loss of prestress is preferable to internal visual and sounding inspection.

Source: Bethany McDonald

Figure 13-3 Typical tools and equipment used in visual and sounding inspections

Source: Bethany McDonald

Figure 13-4 Example of longitudinal crack surrounded by hollow

In addition to identifying indications of advanced loss of prestress, internal inspection can identify circumferential cracking, joints in need of repair, spalling concrete, and other anomalies. Most circumferential cracking is not overly concerning. However, circumferential cracking can be an indication of differential external loads. Differential settlement often produces numerous cracks at the interface between the soil and the structure. Such cracks may be healed or active with visible corrosion products. Internal inspections can also identify where joints have moved slightly and corrosion is occurring or where mortar at joints has failed, allowing corrosion. Periodic identification and repair of cracked joint mortar can protect the joint from internal corrosion. An experienced inspector can identify other anomalies which, if identified early and corrected, may significantly prolong the life of the pipe.

Internal visual and sounding inspection is the only inspection that can identify distress due to longitudinal forces and pipe overload without broken wires. Pipe overload may result in longitudinal cracking near the pipe invert or crown. Inadequate thrust restraint may result in failure of the joint harness or weld, causing joint openings, steel cylinder yielding, cracking of the weld between the steel cylinder and the joint rings, and circumferential cracking of the core.

Visual and sounding deployment platforms and access requirements for internal inspection. Because sounding inspection requires striking the pipe wall with a hammer and listening for the acoustic response, it must be completed during physical entry inspection. Inspection rates vary according to the degree of inspection, pipe diameter, quantity of observations noted, and cleanliness of the pipeline. Typical inspection rates are approximately one to two miles per day. Visual inspection alone can be completed using a robotic inspection platform. Robotic inspection platforms are discussed in Chapter 8. Manned visual and sounding inspections are typically limited to pipes 36 in. in diameter and larger. Pipes must be isolated from the distribution and transmission systems, dewatered, and made safe for manned access. Pipelines smaller than 36 in. can be visually inspected robotically. Robotic inspection platforms are discussed in Chapter 8.

Visual and sounding deployment platforms and access requirements for external inspections. External visual and sounding inspection can be used to verify the results of internal inspections or whenever a pipe is excavated or otherwise available for external inspection. External visual and sounding inspection seeks to find the same evidence of loss of prestress as internal inspection. Loss of prestress that is detected in EM or internal visual and sounding inspections is frequently even more evident during external inspection. External inspection requires exposure of the pipe exterior from the crown to at least the springline. Ideally, the full circumference of the pipe is exposed. The location of spot external locations can be selected from among locations of high consequence, suspected elevated corrosion factors, or easy access.

Visual and sounding inspection limitations. Visual and sounding inspections can be relatively economical, but they do have the following limitations.

- Internal sounding inspections require personnel access to the drained interior of the pipe.
- Pipe surface may be obscured by deposits or biofilms. Water and sediment remaining at the pipe invert will obscure the invert.
- It cannot identify all early stages of deterioration, as it can only identify pipes in the most advanced stages of deterioration, resulting in false negatives for minimally deteriorated pipe. It can also result in false positive identifications.
- It cannot quantify the extent of deterioration, other than the size of hollow-sounding areas and length and width of cracks.
- Experience is required in differentiating between significant and insignificant anomalies.

Electromagnetic (EM) inspection. The EM technology for PCCP wire break detection was developed and patented in the mid-1990s. The first commercial EM tool was subsequently constructed for inspecting in situ PCCP. Multiple variations of the EM technology and its associated electronics on various manned and unmanned platforms have since been developed.

EM inspections ascertain an EM signature for each pipe to identify anomalies that are produced by zones of broken wire wraps. Various characteristics associated with an anomaly (length, magnitude, phase shift, etc.) are evaluated to provide an estimate of the number of broken wire wraps. This inspection method can quantify the amount of wire wrap damage and is the most accepted method of determining the baseline condition of a pipeline. For many years, it has been possible to exploit the concept of eddy currents to measure structural properties in metals. The application of a time-varying magnetic field to metal structures can create internal electric currents as free electrons are driven by the field along discontinuities in the metal itself. Many applications of this phenomenon have been developed to detect damaged sections in steel and iron pipelines (see also Chapter 11).

For PCCP, a different mechanism exists that can be used to determine the structural condition of the pipe. Eddy currents that are generated in a wire wrap can flow along the wire, generating a solenoidal field (Figure 13-5). If the current is interrupted by a break in the wire, the field will be affected. The electromagnetic assessment tool generates currents in the wire wrap and detects where the resulting electric field is altered by breaks in the wire.

To create an electric current in the wire, the condition assessment tool generates an electromagnetic field inside a PCCP. A low-frequency alternating electric current (typically less than 100 Hz) is fed into a coil of wire (known as an exciter coil) positioned near the inner surface of the pipe. The electromagnetic field generated by this coil extends through the concrete core, steel cylinder, and finally into the wire wraps. As the coil travels along

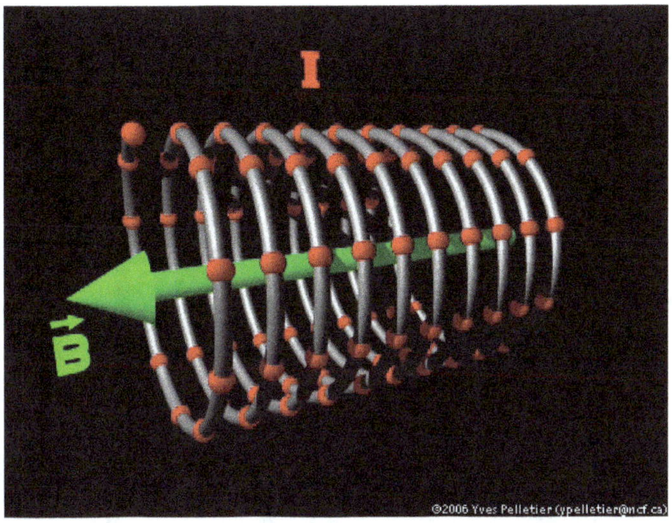

Source: Pure Technologies

Figure 13-5 Electrical currents induced by time-varying magnetic field

the length of the pipe, the field moves as well, creating currents in the nearby wires. As long as the wire is continuous with no breaks, the current will flow uniformly along the wire; however, where a broken wire exists, a discontinuity in the current forms. When the field passes over a section of the broken wire, currents are generated that form opposing magnetic field lines.

Detectors are placed on the opposite side of the pipe from the exciter coil to record the variations in the magnetic field that are created when broken wire wraps interrupt the current flow. Analyzing and interpreting the response of the electrical field allows for estimates of the number of broken wire wraps and the approximate location of the broken wire wraps along the length of the pipe.

EM inspection deployment platforms. The deployment platform is the means for inserting the technology into the pipe and extracting the data from the pipe. Technologies must be selected based on the pipe material being inspected. The deployment platform is selected based on pipeline diameter and access availability. The types of deployment platforms used include manned, tethered robotics, and untethered robotics platforms.

EM inspection of PCCP was first deployed on a manned inspection cart. This platform has the advantage of high dependability and typically lower inspection cost. However, the manned cart requires that the pipeline be fully dewatered and confined space support be provided for personnel entry. Manned EM inspections typically involve one hour for in-pipe equipment setup. Inspection rates vary but typically range from 0.5 to 2 miles per hour, depending on the number of obstacles, diameter changes, and pipe sections encountered containing heavy gauge cylinder. Figure 13-6 shows a technician using a manned EM inspection cart.

Robotics platforms are also available for deployment of EM technology. Using a robotics platform, EM inspection can be conducted by depressurizing a pipeline at high points rather than dewatering the entire pipeline. The tethered robotics platform is tracked and can carry sensors in addition to EM inspection. Tethered EM platforms can inspect up to 9,000 feet in either direction from a single entry point, requiring one to two hours to set up the equipment and progressing at approximately 0.5 to 1 miles per hour. Inspection rates depend primarily on the increasing resistance force in the cable that develops as the robot traverses around bends and over greater distance. Figure 13-7 shows a tethered robotics platform before deployment in a PCCP pipeline.

Source: Pure Technologies

Figure 13-6 Technician pushing an electromagnetic inspection cart through a PCCP

Source: Pure Technologies

Figure 13-7 Tethered and untethered robotics platforms equipped with electromagnetic inspection technology

An untethered robotic platform has been developed that enables EM inspections to be conducted in pipelines that are fully pressurized and in service. This platform uses the flow of water for propulsion and can traverse obstacles including some types of butterfly valves. Untethered robotics platform inspections typically involve a few hours for setup and insertion before progressing at approximately 95% the velocity of flow in the pipeline. Figure 13-7 shows an untethered robotics platform as it is lowered into a depressurized PCCP pipeline.

Electromagnetic inspection access requirements for internal inspection. The access requirements to conduct an internal electromagnetic inspection depend on the platform used. Manned platforms require manned entry, so pipelines must be isolated from the distribution and transmission systems, dewatered, and made safe for manned access. Manned electromagnetic inspection is typically performed only on pipelines 36 in. and larger. For tethered robotics, the minimum access size at a depressurized high point is currently 18 in. Tethered robotics can traverse minor diameter changes and a limited number of bends but cannot traverse butterfly valves. Access requirements for untethered robotics inspections depend on whether the tool insertion and extraction are to be conducted under pressurized conditions. If the pipeline must remain fully pressurized at all times, specialized insertion and extraction tubes that have many requirements and a detailed drawing review are necessary to determine whether access modifications are necessary. If the untethered robotics tool can be inserted and extracted by temporarily isolating and depressurizing at a high point, then the tool can be inserted through an 8-in. or larger access for pipelines up to 36 in. in diameter and 12 in. or larger access for pipelines greater than 36 in. in diameter. The untethered robotics tool can pass some types of butterfly valves and most obstacles on a typical transmission main. Access and pipeline geometry limitations for all platforms continue to change as the platforms develop, so the limitations described above should be confirmed before choosing a method.

Electromagnetic inspection access requirements for external inspection. External electromagnetic inspection tools can be used to confirm the location and quantity of broken wire wraps on excavated prestressed pipes. In order to facilitate this inspection, an excavation is required that exposes at least the top portion of the pipe to be inspected, with at least the top one-fourth of the pipe free of ground cover. In addition, the entire pipe section from joint to joint must be clean enough to allow the tool to traverse the pipe with minimal disturbance. Since the EM coupling with this tool does not need to pass through a steel cylinder, the external EM inspection tool uses a much higher frequency. Ideally, the excavation for an external EM inspection should extend past both joints of the inspected pipe to allow the tool to fully scan both joint sections and identify any damage near the joints. Figure 13-8 shows a pipe excavated for external inspection.

Electromagnetic inspection limitations: EM inspections detect EM anomalies, or differences, in the expected induced field of a PCCP. Anomalies that are consistent with broken wire wraps are of particular importance. However, other interference can mask or distort the electromagnetic signal, affecting the ability to detect and quantify broken wire wraps. This section summarizes factors that can affect the accuracy of detection and quantification of broken wire wraps. Additional detail should be obtained from the inspection provider before performing an inspection.

Accuracy and completeness of the information supplied by the owner: Drawings that indicate how each class of pipe is constructed (cylinder thickness, wire diameter and spacing, shorting strap or non-shorting strap, presence of outlets or other appurtenances, etc.) are used to identify and quantify regions of distress. Discrepancies in the drawings and the data may affect the accuracy of the analysis.

Existence of ferromagnetic (steel) materials near the pipeline: Extra steel near the pipeline can cause a signal distortion that may mask broken wire wraps or cause anomalies that may be misinterpreted as broken wire wraps.

Source: Pure Technologies

Figure 13-8 Example of pipe excavation for external electromagnetic inspection

Proximity to power lines: In some cases, nearby power lines can cause distortion in the signal due to the stray electromagnetic fields. This can limit the effectiveness of the analysis if the distortion is too severe. This interference is rare but is noted for completeness of this manual.

Embedded cylinder pipe without shorting straps: While the EM signal can be very sensitive to a single wire break in this type of pipe, wire-break quantities tend to be overreported. Calibration of the EM signal to actual wire-break quantities is particularly important for this type of pipe, but even with calibration, overreporting of wire-break quantities does occur. Due to these inherent issues with this pipe type, it is recommended verification tests and/or calibrations are performed. In some cases, verification tests can be used as a form of calibration (McDonald et al. 2016).

Wire breaks near a joint: EM is unable to detect wire breaks near joints. The exact distance from a joint in which EM is not able to detect wire breaks varies but has been shown to be as great as 18 in.

Noncontiguous broken wire wraps: When broken wire wraps are separated by unbroken wire wraps, the nonbroken wire wraps can be masked by distress signals and may appear broken depending on the distance from the broken and unbroken prestressed wire wraps. Noncontiguous broken wire wraps may lead to an anomaly that is larger than the actual associated prestressed wire damage.

Verification of EM inspection results for ECP-type PCCP can be accomplished externally using continuity testing, which takes advantage of the fact that a broken wire wrap lacks electrical continuity. Continuity testing requires carefully chipping a strip of mortar, approximately 6 in. wide, to expose wires, typically at the crown of the pipe. A power wire wheel is used to remove surface debris or corrosion from the wires to improve contact with the test leads. Continuity testing measures the differential voltage, also known as potential, between adjacent wire turns using a high-impedance digital multimeter. Measuring the differential voltage between adjacent wire wraps relies on the concept that electrically continuous surfaces are equipotential (i.e., voltage difference between adjacent wire wraps with no breaks is close to zero). However, for a physically broken and separated (electrically discontinuous) wire wrap, a differential voltage between adjacent wire wraps will exist because the surfaces are not equipotential. The potential is measurable because it is affected by multiple, parallel current paths across electrolytic boundaries including the path across the break, the path between adjacent turns through the concrete, and the path through the concrete and the surrounding soil. Because the wire must be the primary electrical path, electrical continuity testing is not applicable to pipe with shorting straps.

Sonic and ultrasonic technology. Sonic and ultrasonic technology has been used for more than 20 years in some large water utilities for condition assessment of PCCP (Fick and Wagner 2010, Fisk and Marshall 2010). Sonic and ultrasonic measurements are conducted to identify flaws such as wire failures, delamination, cracks, and wall thinning in PCCP. In sonic and ultrasonic testing, the strength and cracking of the concrete core are identified by measuring the velocity of compressional and shear waves propagating through the concrete core. The pipe wall thickness, thinning of mortar coating, and existence of delamination between layers of pipe components are identified by measuring the impact echo. Impact echo can also check whether the prestressing wires remain stressed to keep the pipe as a composite structure by evaluating the signal attenuation characteristics (Ge and Sinha 2014, Wardany 2008, Lewis and Fisk 2005).

Sonic and ultrasonic technology can supplement and verify data collected by electromagnetic inspection, especially to identify broken wires near pipe joints, where electromagnetic inspection is generally not effective. Sonic and ultrasonic measurements can evaluate the condition of the compromised concrete at pipe ends due to broken prestressing wires or wires without adequate prestress. It can also be used to detect micro-cracking in pipe coating caused by poor bedding such as pipes sitting on rocks (Marshall et al. 2014). Micro-cracking in the coating could lead to corrosion and breakage of the prestressing wires, which can lead to PCCP failure.

Sonic and ultrasonic technology has some limitations, including the following.

- It requires personnel access to the drained interior of the pipe (Figure 13-9).
- Even though it can be used outside for small-diameter PCCP by exposing the pipe, most inspections are conducted inside the pipe with manned entry. Therefore, it is typically only used for large-diameter pipe.
- The inspection is a manual and slow process, which is time-consuming for long distances.
- This technology cannot determine the number of broken prestressing wires. The minimum number of broken wires required for detection by sonic and ultrasonic technology is not clear.

Indirect Assessments

Indirect assessments are defined as those that identify conditions that can adversely affect PCCP or identify indicators of deteriorated PCCP.

Pressure monitoring and transient mitigation. Like most pipe types, PCCP is typically designed with a surge allowance. However, recurring and high-pressure surges can cause PCCP damage and reduce expected service life. Surges have been identified as the event immediately preceding PCCP failure. Similarly, operating PCCP mains above design pressure can also reduce expected service life. This section describes how pressure monitoring and transient mitigation can be used to extend the useful life of PCCP mains. These methods are applicable to other types of pipes as well, particularly those with brittle fracture characteristics, such as cast iron, asbestos cement, and polyvinyl chloride.

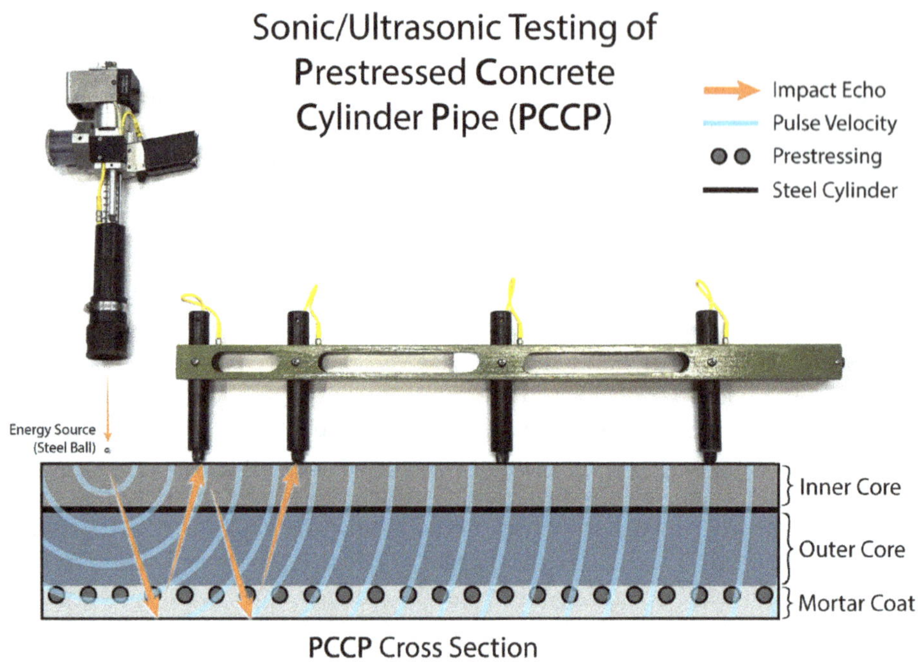

Source: NDT Corporation; created by Katie Roorda

Figure 13-9 Illustration of sonic/ultrasonic testing of PCCP

Pressure monitoring. PCCP mains operated regularly at or above design pressure are more susceptible to developing cracks in the concrete core and external concrete cover, which can initiate corrosion and subsequent breakage of prestressing wires. A good understanding of system operation using pipe design and operating data can allow the owner to determine which pipes operate near or above their design pressures. Typical supervisory control and data acquisition (SCADA) system data is sufficient to determine operating pressure trends over relatively long timespans. SCADA pressure data should be reviewed periodically in comparison to pipe design to identify mains operating near or over design pressure. Pipe design pressure should always be considered when system modifications are made (e.g., booster pump station replacement or addition).

Transient monitoring and mitigation. Transients can occur in any type of pipe. Transients have many sources but are commonly caused or exacerbated by pump on/off cycles, operation of pump control valves, pipeline elevation changes, and in-line valve operation. A typical transient monitor comprises a pressure transducer that is inserted into the main to measure pressure, a data processing and communications unit, and power supply. Selection of a transient monitor is typically based on several factors.

- Pressure sampling frequency: The extent to which a monitor can detect the true peak of a transient depends on the sampling frequency of the unit. Typical SCADA data systems do not collect or store pressure data at frequencies sufficient to understand whether a pipe is experiencing transients or to measure the peak minimum or maximum values. High-speed transient monitors, which measure and store pressures at least 100 times per second, are often needed.
- Data acquisition and notification: Web-based data portals are often used by an owner to view and download data. Real-time notifications can also be received on some types of units.

- Power supply: Transient monitors are usually hard-wired or battery powered. Availability of local power will influence the selection of the power supply.
- SCADA compatibility: Some transient pressure monitors can also link to the owner's SCADA system.
- Need for other sensors: Additional sensors for pressure and flow can be connected to some types of transient monitors.

Transients can propagate long distances through pipelines, simplifying detection of the presence of transients but complicating troubleshooting the source of the transient. Locating the cause of a pressure transient may require installation of multiple monitors to bracket or narrow down the source of the transient.

Acoustic wire break monitoring. Acoustic monitoring of PCCP is a technique for recording and locating prestressing wire breaks in operational pipelines. The technology acoustically detects wire breaks within operating PCCP as they occur. The acoustic energy associated with a PCCP wire break can be detected and identified. Data acquisition and processing equipment and software estimate the location and time the wire break occurred.

Acoustic wire break detection was initially applied to PCCP in the late 1990s using autonomous sensors. The initial sensors were accelerometer acoustic sensors mounted to the surface of the pipe or hydrophones installed in the water column via a small-diameter tap or corporation stop. Locating the source of the wire breaks relied on bracketing the acoustic activity associated with the wire break between two sensors. The location of the wire break was estimated based on the strength of the acoustic signal detected by each sensor or hydrophone.

In 2006, acoustic monitoring using a distributed fiber optic platform was introduced. A fiber optic cable is acoustically sensitive throughout its entire length. Because the sensor is never more than one pipe diameter away from a wire break, acoustic detection, signal resolution, and wire break–locating capability are better than with paired, autonomous sensors.

Acoustic wire break monitoring deployment platforms. Acoustic monitoring of PCCP for wire breaks is currently available using a fiber optic deployment platform which is installed in the full length of the monitored pipe. Internet-based data reporting is provided.

Acoustic wire break monitoring access requirements. Fiber optic acoustic wire break monitoring can be installed while the pipeline is dewatered (dry deployment) or under normal flow and pressure (wet deployment). The dry deployment consists of unspooling the fiber optic sensor along the invert of the pipeline and periodically fastening the fiber to the pipe. Alternatively, the fiber can be installed while the main is in operation. For wet deployment, a fiber optic sensor cable can be deployed into an operational pipeline with limited flow velocity (<2 ft/s). Once the fiber optic sensor is secured at both ends of the pipeline, data acquisition equipment is connected to one end of the fiber and the pipeline is monitored. Figure 13-10 shows both deployment types.

Limitations of acoustic wire break monitoring include the following.

- Wire breaks that occur before the system is installed are not detected by this system. This is known as the "baseline" condition, and an electromagnetic inspection just before system installation should be completed to understand the baseline condition.
- Wire breaks that occur when the detection system is shut down for maintenance or when an extended power outage occurs are not recorded. Since the system is typically equipped with an uninterruptible power system, this limitation is often negligible.
- This system monitors the pipeline for breaking prestress wires, which is the predominant failure mechanism of PCCP. In some instances, PCCP may fail in ways that do not involve the progressive deterioration of the prestressed wires. This system will help to mitigate failures not associated with wire breaks.

Source: Pure Technologies

Figure 13-10 Fiber optic cable anchored to pipe in dry deployment (left) and deployed into in-service pipeline in wet deployment (right)

- Future pipe modifications will require modification of the fiber. For example, if a distressed pipe section is removed and replaced, as is common in PCCP management, the cable will need to be cut during PCCP removal, protected during installation of the new pipe, and spliced before the main is returned to service.
- Installation through pipe with obstructions requires specialized design. For example, installing fiber in a main with in-line valves typically requires the fiber to be routed outside the pipe, around the valve, and back into the pipe on the other side of the valve.
- While wire break monitoring is an acoustic technology, it cannot currently detect leaks within the pipe.

Corrosivity surveys. Corrosivity surveys can be used to identify soil and groundwater conditions that may be deleterious to PCCP and to identify locations of active corrosion of the pipe. Refer to Chapter 5 for descriptions of these tests.

Acoustic leak detection. As with any pipe, leaks can occur on PCCP. The applicability of leak detection as a viable condition assessment tool depends on whether leakage is a typical failure mode or an indicator of a typical failure mode of the pipe. For PCCP, owners generally describe sudden failure with no reported leaks immediately preceding the failure. This failure mode is supported by the fact that prestressing wires are known to be the primary structural component of PCCP, and breakage of the prestressing wires can result in sudden failure of the pipe without a breach of the steel cylinder or concrete core. Generally, a large area of weakness (i.e., a large number of broken wires) must exist before the concrete core and steel cylinder rupture—hence the catastrophic nature of PCCP failures.

Leak detection in pipelines is described in detail in Chapter 6. Leak detection can be applied to PCCP but is generally not accepted as a means of identifying PCCP in danger of failure.

STRATEGIES FOR MANAGING DISTRESSED PCCP

PCCP is a pipe that requires a strategy to maintain and manage it. The more diligent the strategy, the fewer risks the owner assumes. Several strategies are commonly used by owners to protect their PCCP and manage the risk of failure. Some of these strategies are discussed here.

- Protect, inspect, and monitor
- Protect, inspect, evaluate, and repair or systematically replace

The components of these strategies are explained in this section.

Protect

Protecting PCCP is similar to protecting any pipeline. The owner wants to protect the pipe from being overloaded, prevent damage by third-party contractors, maintain accurate records of the pipe's location, and minimize surge events, etc. However, protection of PCCP should also include protecting it from stray currents.

An owner should regularly perform a stray-current assessment of the PCCP. This assessment should focus on areas where "stray-current interference" is affecting the PCCP. Stray-current interference occurs when DC electrical current flowing through the ground "strays" from its intended path onto a foreign structure (a PCCP pipeline in this case). Stray-current interference occurs because a PCCP pipeline has a lower resistance than the ground. The pipe provides an easier path for the current to travel upon, so it strays from its intended location. Eventually, the stray current discharges from the PCCP back into the ground and at these locations corrosion occurs on the PCCP pipeline. A stray-current survey looks for locations where a significant amount of stray current is affecting a PCCP pipeline. Once stray current is detected, the owner will have to decide how to protect the pipe. Options to handle this issue include the following.

- **Install an impressed-current cathodic protection system:** An impressed-current cathodic protection system will prevent both general corrosion and stray-current corrosion. If an impressed-current system is selected, the owner must design the system carefully because if it is done improperly it can damage the PCCP. Damage can occur in the PCCP because the high-strength steel wires in PCCP are susceptible to damage from hydrogen embrittlement when impressed-current cathodic protection is used.
- **Install a galvanic cathodic protection system:** A galvanic cathodic protection system will prevent both general corrosion and stray-current corrosion. The system uses a sacrificial anode (usually made of magnesium or zinc), which is "sacrificed" to provide protective current. Galvanic protection provides enough current to effectively protect the PCCP pipeline but with little risk of embrittling the PCCP high-strength wires because the sacrificial anodes do not produce enough current to cause embrittlement. Because anodes are consumed over time, they must be periodically monitored and replaced when needed.
- **Bonding:** Wires can be installed between the affected and offending pipelines to electrically "bond" the pipelines and provide a path for the current to return to its source without discharging into the ground. However, there remains a risk of hydrogen embrittlement at the bond location, so caution must be exercised.
- **Sacrificial anode stray-current drains:** Sacrificial anode stray-current drains provide a path for stray current to discharge or "drain" from the pipeline back to the ground. Instead of causing corrosion to the pipeline, the anodes are consumed and can be replaced when depleted. Sacrificial cathodic protection anodes are used in this method, but the pipeline is not considered cathodically protected, because only enough anodes are employed to drain the stray current, not to cathodically protect the pipe.

Inspect

The inspection component of the strategy consists of more than just inspecting the pipeline periodically. The following issues should be considered when developing an inspection program.

- **Prioritization:** The owner needs to prioritize the pipelines in their system so that the pipeline(s) with the greatest risk are inspected first. The factors considered in any prioritization will depend on the owner risk tolerance and preferences. Common factors include: (1) review of pipeline age and history, including wire type and manufacturer; (2) evaluation of the number and severity of existing wire breaks; (3) assessment of the pipeline itself (considering factors such as the design, construction, fabrication, as-built conditions, repair history, pipe operating conditions, and pipe pressures); (4) evaluation of environmental deterioration factors (soil corrosivity and third-party stray currents); and (5) additional factors such as redundancy and agreements with other agencies that may be served by the pipeline.
- **System operation:** Consider the impact of the system's operation on the pipeline. Factors to consider include surge events, operating pressure in relation to design pressure, pressure fluctuations, and increases in loading conditions over time. This evaluation should also consider the most feasible way to conduct the inspection and the impacts of taking the line out of service. These impacts should be communicated to the affected customers.
- **Conduct EM inspections:** Currently EM inspection is the industry's primary means for identifying wire breaks. The industry currently has a single provider of this service. The owner will also have to provide the logistical support needed to conduct the inspection.
- **Conduct visual and sounding inspections:** Visual and sounding inspections can be conducted along with EM inspections to identify joint issues and other damage not detectable using EM technology.
- **Evaluate the results:** After conducting the inspections, the owner needs to evaluate the results and develop a strategy to deal with the pipe sections with broken wires. The results can be plotted on a geographic information system, which is a visual method to see the degree and amount of pipe degradation (see Chapter 16).
- **Schedule next inspection:** Based on the results of the inspection, the owner should determine the interval between inspections. Inspections should be scheduled between 5 and 10 years apart depending on the pipe's characteristics and degradation of the pipe segments.
- **Track condition over time:** In some cases, the rate of deterioration is not consistent over time. If the primary source of the damage (e.g., transient pressure surges) is removed, the rate of deterioration may slow significantly. On the other hand, wire breakage can lead to more cracking and acceleration of corrosion. The management strategy required will vary according to the rate of deterioration.

Monitor

Monitoring the pipeline involves installing an acoustic fiber optic monitoring system. This system provides real-time detection of wire breaks. The system consists of a fiber optic cable laid inside the pipe and a data acquisition unit where the data are recorded and evaluated. The cable acts as a sensor that "listens" for prestressing wire breaks while the pipeline is in service. The location of wire breaks is calculated from sensor locations based on the speed of sound, so once the sensors are installed, information is collected on a continuous real-time basis. This system, when combined with a baseline inspection, will allow owners to know with relatively good accuracy how many wire breaks are in any particular pipe section at any time. The system can be configured so that "wire break alerts" can be sent directly to the owner immediately after a wire break, allowing the owner to respond appropriately.

Evaluate

After inspecting a PCCP pipeline with an area of broken wires, the owner must decide what to do with the pipe. This is a difficult decision to make without the technical orientation to determine how wire breaks affect pipe performance. To help make this decision, this section outlines several options used by owners.

Wire break limit estimation. In this case, the owner can rely on any of several methods to determine the maximum allowable wire breaks. These methods typically include engaging internal staff or an external consultant to develop an estimate based on past PCCP experience. This decision should be based on the internal pressure of the pipe and the number of surge events the pipe experiences. Once this limit is reached, the pipe should be repaired.

Failure margin analysis using risk curves. This method evaluates the effect of broken prestressing wires on the performance of the pipe and its margin to failure using a calibrated and verified model. Failure margin may be evaluated using risk curves corresponding to limit states related to serviceability, damage, and ultimate strength of the pipe with broken prestressing wires. The failure risk analysis provides for each distressed pipe a relationship between the number of broken wires and the internal pressure that causes the certain limit states to be exceeded. A thorough summary of this approach is provided in Zarghamee et al. 2012.

Serviceability limit estimation. In this approach, the pipeline serviceability limit is estimated for no wire breaks and relatively low quantities of wire breaks using the design equations presented in C304, "Design of Prestressed Concrete Pressure Pipe," (2014) in a spreadsheet-based risk-analysis tool. Risk is presented in the form of curves that show how pipe pressure capacity decreases with increasing numbers of wire breaks. The model estimates the effect of wire breaks as a loss of prestressing strength over the length of the pipe. This approach was reported by Rajah et at. in 2014.

After determining how "bad" the pipe is, the owner needs to develop a strategy for dealing with the PCCP in the future. Three options have commonly been used.

- *Live with a degraded pipe and reinspect it in a few years.* This option means that the owner does not believe the broken wires have weakened the pipe sufficiently to warrant repairing the pipe. This philosophy is commonly used and is justifiable provided the owner is committed to reinspection within a few years. In this way, the pipe's degradation can be monitored and the pipe can be repaired before it fails. To determine the proper inspection schedule, the owner will need to consider the corrosivity of the environment, the potential for stray currents and hydrogen embrittlement, the operating conditions, and the history of the pipeline. At this time, the owner should also consider whether the main or individual pipes should be scheduled or budgeted for long-term repair or replacement.
- *Install a system that will monitor the pipe's condition so it can be repaired when the pipe degradation is serious.* This method involves installing an acoustic fiber optic monitoring system. This system combined with an initial inspection will tell the owner the state of the pipe condition in real time. This method allows the owner to accurately monitor the pipe's degradation. When the degradation is serious enough or if the pipeline wire break rate grows quickly, the owner can repair or replace it.

- *Repair or replace the pipe section(s) with wire breaks.* This method involves repairing or replacing the pipe that has an unacceptable number of wire breaks, restoring its original strength. The project many involve a single pipe section or multiple contiguous or noncontiguous pipe sections, depending on the owner's budget, risk tolerance, and operational flexibility.

Managing PCCP involves difficult decisions that must be made by utility management and should be based on many factors, including the risk and consequences of failure, budget, and prioritization of repairs with management of other infrastructure in the system.

Rehabilitation and Replacements

Owners use two strategies to rehabilitate and replace PCCP. The first is to rehabilitate or replacement individual distressed pipe sections, and the other is to rehabilitate or replace the entire PCCP pipeline. These strategies are explained here.

Rehabilitation or replacement of individual distressed segments. In this approach, the owner prioritizes the distressed pipe sections and addresses the PCCP segments that have an unacceptably high risk of failure. The owner plans and implements the rehabilitation or replacement based on the risk associated with the distress. This approach only addresses the pipe sections in distress. This approach minimizes capital expenditures and reduces the number and length of service interruptions. Efforts may entail replacing the pipe, lining the pipe with carbon-fiber reinforced polymer, lining the pipe with steel, or using a hybrid, proprietary system.

Rehabilitation or replacement of an entire pipeline. Owners who use this more conservative approach believe it is more cost-effective to rehabilitate or replace an entire pipeline rather than addressing individual damaged sections. These owners generally have a large number of pipe sections with wire breaks in relative close proximity, so rehabilitating or replacing a longer reach of the pipeline is more cost-effective than making repairs in the same area every few years. This is especially true when the owner expects that additional pipe sections will show distress over time.

The goal of this strategy is to reduce the risk of PCCP failure and reduce the cost of multiple individual repairs. This approach restores the remaining useful life of an entire segment of the PCCP to a new design life, based on the design life of the replacement. In addition, while making these modifications the owner can also update additional aspects of the line (such as valves, meters, manholes, and other appurtenant structures). This approach requires significantly more capital, longer shutdowns, and a multiyear repair schedule.

REFERENCES

ANSI/AWWA (American National Standards Institute/American Water Works Association). 2014. C301. Prestressed Concrete Pressure Pipe, Steel-Cylinder Type, for Water and Other Liquids. Denver: AWWA.

ANSI/AWWA. 2014. C304. Design of Prestressed Concrete Pressure Pipe. Denver: AWWA.

AWWA. M9, *Concrete Pressure Pipe*. 2008 (3rd ed.). Denver: AWWA.

Fick, G. and Wagner, T.B. 2010. Leading the Way: The Washington Suburban Sanitary Commission's Comprehensive PCCP Management Program. American Society of Civil Engineers. Proc. Pipeline Division Specialty Conference, Keystone, Colo.

Fisk, P.S. and Marshall, J. 2010. Detecting Deteriorating Thinning PCCP Mortar Coating. American Society of Civil Engineers. Proc. Pipeline Division Specialty Conference, Keystone, Colo.

Ge, S. and Sinha, S. 2014. Failure Analysis, Condition Assessment Technologies, and Performance Prediction of Prestressed-Concrete Cylinder Pipe: State-of-the-Art Literature Review. *Journal of Performance of Constructed Facilities*, 28(3):618–628.

Lewis, R.A. and Fisk, P. 2005. Detection of prestressed concrete cylinder pipe thinning from hydrogen sulfide deterioration. American Society of Civil Engineers. Proc. Pipeline Division Specialty Conference, Houston.

Marshall, J.; Fisk, P.S.; and Fontaine, L.E. 2014. Detecting the Effects of Inadequate PCCP Bedding. Proc. Pipelines, Portland, Ore.

McDonald, B.A.; Riggs, R.; and Conroy, A. 2016. Using Electrical Continuity Testing to Improve Electromagnetic Inspection Results. American Society of Civil Engineers. Proc. ASCE Pipelines Conference, Kansas City, Mo.

Rajah, S.; Plattsmier, J.; McDonald, B.; and Conroy, A.D. 2014. Development of a Risk Assessment Tool for Managing PCCP Assets with Broken Wires. American Society of Civil Engineers. Proc. ASCE Pipelines Conference, Portland, Ore.

Romer, A.E.; Ellison, D.; Bell, G.E.C.; and Clark, B. 2008. Failure of Prestressed Concrete Cylinder Pipe. http://www.waterrf.org/PublicReportLibrary/91214.pdf (accessed Dec. 17, 2018). Denver: Water Research Foundation.

Wardany, R.A. 2008. Condition Assessment of Prestressed Concrete Cylinder Water Pipes. Proc. 60th Annual Western Canada Water Annual Conference & Trade Show, Regina, Saskatchewan.

Zarghamee, M. S.; Ojdrovic, R.P.; and Nardini, P.D. 2012. *Best Practices Manual for Prestressed Concrete Pipe Condition Assessment: What Works? What Doesn't? What's Next?* Denver: Water Research Foundation.

This page intentionally blank.

Chapter 14

Hydrostatic (Pressure) Testing of Existing Pipes

Andrew E. Romer, Chapter Lead, AECOM
Dan Ellison, HDR
Michael Grahek, Los Angeles Department of Water & Power

Hydrostatic examination, also known as pressure testing, is a quick and inexpensive method that can be used to find weak areas of pipe—those that could produce a large rupture. Unlike the spot assessment techniques mentioned elsewhere in this manual, hydrostatic examination can test the entire pipeline. The process involves pressure testing sections of pipeline so that any weak areas can be detected and further assessed in a controlled manner. This can be done with lower risk of property or other consequential damage than waiting for the pipeline to fail. If no weak areas are discovered, the owner is provided assurance that the pipeline will continue to perform well without an immediate risk of a catastrophic break.

Although this is both the oldest and the most universally applied technique for testing new pipes, pressure testing has not often been applied to old water pipes. Perhaps this is attributable to a few conceptual stumbling blocks. Testing new pipe is different from testing old pipe. A pressure test of a new pipe can be considered primarily a "leak" test, whereas a test of an old pipe is intended to be a "strength" test. The material integrity of the new pipe is seldom in question, so tests focus primarily on whether the joints are sealed and adequately restrained. On an old pipe, integrity of the material is in question, and the focus is on seeing if the pipe will not fail when subject to a higher stress than seen under typical operating conditions.

Trying to fail pipes to save them may be a rather difficult concept for many to accept. Although a leak test of an old pipe might be desirable, it may not be feasible. Achieving a tightly closed system may be too difficult, with old valves and numerous service connections providing too many avenues for water to escape. In fact, pressure testing of old pipe could almost be described as a "hydrodynamic" rather than a "hydrostatic" test, with a large pump needed to overcome the leakage past the closed valves. The result is that when a pressure test of an old pipe is completed, the pipe may not be declared "leak-free," but it may be sufficient to attest to its continued serviceability.

In some publications (such as those listed at the end of this chapter), the testing of old water mains has been described as "controlled destructive examination." The "controlled" part of the description refers to several things. If the pipe ends up with a break during the test, the risk of flooding and ancillary damage is virtually eliminated. The quantity of flow is strictly limited, and with personnel on the job, water can be shut off as soon as a problem is discerned. Customer service interruptions also are scheduled in advance, and minimized through the use of bypass piping, where needed. Because the potential break is "planned," materials and tools should be readied in advance to minimize the time required to make repairs. Forcing a pipe to break on your schedule is far better than responding to a crisis in the middle of the night.

This method would likely detect pipeline segments that are about to fail and could minimize the problems associated with failure. However, a "successful" test would not necessarily indicate an absence of bad material, and a pipeline that passed testing could still fail at some point in the future. Plastic materials often take long periods of time (days, months, or years) before they strain to the failure point. A pressure test that is limited to several hours may not achieve the necessary strain. Metal pipes will also continue to corrode, eventually leading to future failures. If a metal pipe can endure a significantly elevated pressure today, the owner can be confident that future failures will be either (1) modest in size or (2) several years in the future. To maintain such confidence, pressure testing may need to be repeated regularly.

The Los Angeles Department of Water and Power (LADWP) used this method several years ago to evaluate more than 100 miles of large-diameter riveted and welded steel pipe originally installed between 1890 and 1940. Through these tests, the LADWP confirmed that the likelihood of catastrophic failures of these pipelines would be minimal over the next several years, allowing the utility to defer several hundred million dollars of planned replacements.

THE BASICS

Hydrostatic examination entails isolating the main using existing valves, then elevating the pressure using a pump. (Because valves on most large mains do not always close tightly, a large, continuously running pump may be needed.) The pressure is raised in small increments, following an engineer-developed plan, while closely monitoring pressure and flow from the pump. If the pressure suddenly drops and the flow suddenly increases, a leak has been triggered (or a pre-existing leak magnified to a detectable level), and the pump is shut off. Various methods (as described elsewhere in this manual) can then be used to find the leak. During the repair of the leak, the condition of the pipe is assessed at a point of known weakness.

Admittedly, this is counterintuitive—who wants to make a pipe leak? But the elevated pressure is not causing the leak; the leak or potential break is already there, caused by corrosion or other condition problems. The test is merely triggering a leak that will occur eventually (probably at a much less convenient time). Pressure testing is similar to performing a stress test in a doctor's office to detect heart problems—something that could be quite deadly in a less forgiving setting.

Pressure testing is the only practical method that could have found this problem in advance.

Figure 14-1 Catastrophic break due to bad weld at intersection between 30-in. and 36-in. pipes

HYDROSTATIC EXAMINATION

The major drawback of in-pipe nondestructive examination (NDE) is cost. The inspection tool and technical data processing can be expensive, and getting the tool into and out of the pipe can be difficult. For high-resolution assessments, an NDE inspection tool must be sized to match the pipe, so a large tool is needed for a large pipe, requiring a large access port. For some operating pipelines, this may not be practical. Moreover, NDE seldom tests 100 percent of the main and is not capable of finding certain defects, like bad joint welds (Figure 14-1) that may lead to catastrophic breaks. The primary advantage of the hydrostatic test method is that it is much more economical, particularly if few or no problems are detected, and it tests 100 percent of the pipeline segments. Hydrostatic testing involves a temporary shutdown of the pipeline—isolation of a section of the main using existing line valves and service valves. That is followed by pressurization of the isolated main using a large pump. (Some leakage past the closed valves is expected.) Constant monitoring of pressures and flows is important during the test.

The use of leak detection equipment/techniques is appropriate in conjunction with the test should a leak be detected or rupture initiated. And, of course, the pipeline owner should be prepared to dewater and repair the pipeline should a leak be detected. In the case of a detected leak, the utility needs to fix any failure and repeat the test until no additional failures are detected. Fixing the failures would likely involve setting up traffic control, excavating an access hole, repairing the pipe, backfilling, and repaving; however, traffic control is likely already established for the test (at least at the point of pressure connection.)

The cost of hydrostatic examination can vary greatly depending on how much pipeline is tested at one time and whether or not leaks are initiated, but in general, the cost is quite low. This is particularly true if the leak repairs are considered "fixed costs"—something the utility would have borne anyway.

The test itself relies on a strict protocol designed to control events and detect the leaks and breaks. Pressures and flows are controlled and monitored through several steps over an extended time. When a pipe fails during the test, the failure is usually signaled by a sharp drop in pressure and an increase in flow. The pressure testing procedure also calls for active leak detection. Unfortunately, because it is impossible to know how much water is leaking past the closed valves, it is not always possible to tell whether the pipe has a small leak. But discovering small leaks is not the purpose of the test. The goal is to uncover the large areas of weakness that might lead to catastrophic results if allowed to fail on their own. A utility will discover some leaks using pressure testing, but not necessarily all of them.

Application of Hydrostatic Examination

The applicability of the recommended testing methods depends on the prevalence and dispersion of the defective material, as described in Table 14-1.

PROCEDURE DEVELOPMENT

Pressure testing (Figure 14-2) is the oldest and most widely practiced method of testing pressure pipe. In the water industry, it is commonly used for acceptance testing of new pipe installations. In the fuels industry, periodic testing of existing pipe is required by law. In conventional pressure tests, the pipe section is tightly isolated at each boundary and leaks are signaled by a drop in static pressure within the closed system. The isolation is achieved using bulk heads, test seals, or tight-closing valves. For tests of existing water systems, such tight isolation is usually not practical. Operational demands generally will not allow sufficient time for installing and removing large bulkheads in addition to the time required for testing, repair, and (occasionally) retesting. Also, the cost of bulkhead installation could easily triple the cost of a test. As for "tight-closing" valves, those are quite rare in an existing water system.

Table 14-1 Dispersion scenario and appropriate testing method

Prevalence/Dispersion of Defective Material	Appropriate Testing Method
Scenario 1 A very small amount of defective pipe exists	Annual hydrostatic testing of the pipeline would be appropriate to provide confidence in the continued serviceability of the pipeline.
Scenario 2 A large amount of defective pipe exists	A few extracted and tested samples should be able to confirm this. Testing would be followed by rehabilitation or replacement of the entire pipeline.
Scenario 3 The defective material is limited, and confined to the immediate vicinity	The testing needs to find the limits of the defective material. This would be done by starting at the extreme ends of the suspected material and working inward. Rehabilitation or replacement of the pipeline would be done within the limits determined by the tests.
Scenario 4 The defective material is limited, but dispersed throughout	This scenario is the most difficult to address. By testing representative samples of the material, an overall risk of pipeline failure can be estimated. This estimate would then be considered in conjunction with other risk factors to determine the extent of a recommended rehabilitation/replacement project. Among the other risk factors to be considered are (1) the varying pressures along the alignment, (2) the cost of performing additional testing, (3) the cost of pipeline renewal, and (4) the costs and social impacts related to unexpected, uncontrolled failures.

Source: Mike Grahek

Figure 14-2 Pressure testing for LADWP

Thus the "trick" is to use a non-static test procedure that can find leaks in the pipe, tolerate the leaks past the valves, and simultaneously keep flow under strict control.

How do you find leaks? Here again the solution may be "low-tech," i.e. wait and see. LADWP's experience is that most leaks eventually come to the surface, and its hydrostatic examination test specifications require a 72-hour "cure" period after the test is over, in which the pipe is held under normal pressure but restricted flow, giving leaks time to develop. The test engineer also looks for changes in flow and pressure that signal the start of a leak and compares before-and-after flows at various pressures. With these techniques, LADWP has been able to detect all but the smallest of leaks.

Some small leaks may go undetected. Weeks after completing a test, the possibility exists of discovering a modest leak that may have started as a small leak during the tests. This is one of the risks of the methodology and is no cause for embarrassment: the test did not cause the leak; corrosion did, and the leak would have eventually been discovered anyway but likely when the utility was not onsite and prepared to fix it. Remember that finding such small leaks is not the goal of the test. The goal is to find large areas of weakness, those "paper-thin" sections that endanger people and property and cannot stand even 10 percent more pressure.

The in-pipe leak detection methods described in Chapter 7 of this manual are worthy of consideration as an adjunct to the hydrostatic testing described in this chapter. Acoustic leak detection performed before hydrostatic testing may be of value, as would acoustic monitoring during hydrostatic testing. There are numerous acoustic leak detection devices and methodologies, as discussed in Chapter 7. Because the frequencies of sound generated by leaks through pipe walls vary considerably based on the material of the pipeline, its length, and its diameter, an acoustic specialist should be consulted to determine whether that technology is appropriate for any water main.

PRESSURE LIMITS

The pressure applied should not exceed the design pressure or pressure rating of any fitting, valve, or thrust restraint system used on the pipeline (Table 14-2). It should also not exceed the highest previous hydrostatic test pressure. Structural analysis of the components should be conducted before testing in order to confirm that the pipeline had (at least when new) the structural capacity to withstand the test pressure. For example, it would be prudent to analyze the pressure limits of the riveted joints as described in ASCE Manual M79.

Table 14-2 Test pressure limits

Pipeline Material	Maximum Test Pressure	Source
Asbestos cement pipe	2 times operating pressure < the pressure class + 50 psi for sizes 16 in. and smaller. Operating pressure + 50 psi for sizes >16 in.	AWWA C400, C401, C402, C403, and C603
Reinforced concrete pressure pipe—steel cylinder type	120 percent of maximum working pressure	AWWA C300 and M9
Prestressed concrete cylinder pipe—lined cylinder type	120 percent of maximum working pressure	AWWA C301, C304, and M9
Prestressed concrete cylinder pipe—embedded cylinder type	120 percent of maximum working pressure	AWWA C301, C304, and M9
Reinforced concrete pressure pipe—non-cylinder type	120 percent of maximum working pressure	AWWA C302 and M9
Concrete pressure pipe, barwrapped—steel cylinder type	120 percent of maximum working pressure	AWWA C303 and M9
Ductile-iron pipe	1.5 times working pressure or pressure class, whichever is less	AWWA C151, C600, and M41
Fiberglass pressure pipe	*See Note 1*	AWWA M45 and C950
Polyethylene pressure pipe	1.5 times working pressure or 1.5 times rated pressure of pipe at 0.50 design factor whichever is less	AWWA M55 and C906
Polyvinyl chloride (PVC) pressure pipe	120 percent of maximum working pressure	AWWA C900, C605, and M23
Molecularly oriented polyvinyl chloride (PVCO) pressure pipe	120 percent of maximum working pressure. *See Note 2*	AWWA M23 and C909
Wood stave pipe	Not recommended	*See Note 3*
Riveted steel water pipe	*See Note 4*	ASCE Manuals 79 and M11
Welded steel water pipe	125 percent of working pressure	AWWA C200, C206, C604, and M11
Stainless steel water pipe	125 of working pressure	AWWA C220 and M11

Note 1: ASTM D2992 is used to determine the hydrostatic design basis, which is extrapolated on the basis of time-to-failure tests. It is appropriate to determine where on the 50-year nominal time-to-failure curve the pipeline is and limit the test pressure to a value below the hoop stress at which the nominal factor of safety is maintained.

Note 2: Some pipe may not have been previously hydrostatically proof tested. See AWWA C909, Sec. 6.1.2.

Note 3: Consult ASCE *Guidelines for Inspection and Monitoring of In-Service Penstocks* (ASCE 1998) by the Task Committee on Inspection and Monitoring of In-Service Penstocks (Chapter 7) and Section 5.1 of *Guidelines for Evaluating Aging Penstocks* (ASCE 1995).

Note 4: The ASCE *Guidelines for Evaluating Aging Penstocks* (ASCE 1995) indicated that "Field hydrostatic testing of the finished configuration did not appear to be universally required" (pp. 3–14). ASCE Manual 79 *Steel Penstocks* indicates that field test pressure if conducted may have been from 1.1 to 1.5 times the "working pressure" (p. 269). John Bambei, chair of the AWWA Steel Pipe Committee, reported that Denver Water's riveted penstocks were tested to the static Hydraulic Grade Line, so testing under this protocol should not exceed that limit.

In some instances, it may be appropriate to expose pipe or fittings for more careful examination for damage during the test. When testing against closed valves, it is not reasonable to expect that there will be no leakage through the closed valve seats. AWWA C604 (2011), Section 5.2.5, suggests that in this instance, an allowance of 0.0078 gph/in. of nominal valve size (for each valve) should be allowed in addition to makeup water added to maintain pressure within 5 psi of the test pressure. This is for a new valve that seats relatively well. Old valves, particularly those that have not been exercised regularly, will leak considerably more.

Test periods should be short. Test most pipes for no more than 4 h. Shorter test periods may be appropriate. Consult with the source documents listed in Table 14-2.

AFTER THE INITIAL TESTS

For any hydrostatic examination program to succeed, it must become institutionalized. Following an initial round of testing and rehabilitation, there should be retests on a regular schedule. Such a commitment should not be set aside to meet other pressing needs, fall victim to the budget, or be forgotten when its advocates depart. The cost-effectiveness and public responsibility of this kind of program is significant. When the potential property damage and risks to people are considered, the benefits of hydrostatic examination become clear.

REFERENCES

ANSI/AWWA (American National Standards Institute/American Water Works Association). 2011. C604. Installation of Buried Steel Water Pipe - 4 In. (100 mm) and Larger. Denver: AWWA.

ASCE (American Society of Civil Engineers). 2012. Manual of Practice No. 79. Steel Penstocks (MOP 79). Reston, Va.: American Society of Civil Engineers.

Ellison, D.; Bell, G.; Rieber, S.; Spencer, D.; Romer, A.; Matthews, J.; Sterling, R.; and Ariartnum, S. 2014. Answers to Challenging Infrastructure Management Questions. Denver: Water Research Foundation.

Ellison, D.; Romer, A.; Bell, G.; and O'Brien, A. 2001. Distribution Infrastructure Management: Answers to Common Questions. Denver: AwwaRF.

This page intentionally blank.

Chapter 15

Strategies for Economical Assessments of Low-Value Pipes

Dan Ellison, Chapter Lead, HDR
Frank Blaha, Water Research Foundation
Dave Spencer, HDR
Piero Salvo, GAME
Allison Stroebele, Pure
John Marciszewski, Echologics, a division of Mueller

Water infrastructure managers have often remarked that money spent on condition assessment is better spent on main replacement. This is particularly true if the pipe is small and relatively inexpensive to replace. Yet few wastewater utility managers say the same thing. Why the difference? Condition assessment of gravity sewer pipes is simple, low-risk, and relatively inexpensive. Video cameras installed through manholes can often distinguish good pipes from bad pipes, providing pictures that even a layperson can understand. For water systems, the collection of condition assessment information is not so simple, with higher risks, higher unit costs, and only abstract pictures produced. Yet there is considerable evidence that water utilities could save money if condition assessment of low-value mains were more routinely practiced.

This chapter discusses strategies that can be and have been employed on low-value mains as ways of producing benefits that well exceed the cost of the assessments. Greater adoption of these methods would produce a virtuous cycle; a larger market produces more competition and more innovation, driving down costs and creating more value. With more and better data, our ability to generate useful information from condition assessment would improve.

NONDISRUPTIVE SAMPLING METHODS

Common practice in water utilities is to select a main for replacement after it has failed repeatedly within a short period of time. While this time-honored method works in many (if not most) circumstances, there are two potential flaws in this approach.

(1) Good pipe is often discarded as well as bad pipe. This happens when the breaks stem from corrosion pitting and other defects that are relatively isolated and the remainder of the pipe is in relatively good condition.

(2) Breaks occur that could have been avoided. When a condition assessment shows that the pipe was is very poor condition, there is no reason to wait for several breaks to occur before replacing the main. Such breaks disproportionately damage the utility's reputation and waste money.

Selecting Sample Locations

The key to avoiding these flaws is determining whether deterioration is isolated or widespread. An assessment is needed of both the interior and exterior of the pipe to make such a determination. Exposure of the main in random locations for external assessments can be helpful, but a better approach is usually to target areas of the main (particularly if the main is ferrous), where defects are more likely to be found. Random samples can provide a false sense of security. The absence of defects does not necessarily indicate a low likelihood of failure. The exception is when the sampling is part of a well-structured statistical study. The corrosivity survey and leak detection methods described in Chapters 5 and 7, respectively, can provide guidance regarding locations where excavations will be most fruitful.

Sample Extraction

Samples of pipes can be extracted for evaluation. To avoid disruption of operations, coupons have often been taken using hot-tapping techniques. Unfortunately, research has shown that these coupons are not as representative of the pipe as a whole ring section, as corrosion is often more prevalent on the pipe invert or pipe crown, but extraction of a complete ring requires a shutdown and therefore is not always preferred. Once extracted, various spot assessments, as described in Chapter 6, can be applied.

Keyhole Methods

Figure 15-1 shows an alternative approach in which a wraparound scanning tool is employed in a vacuum-excavated "keyhole" excavation. This tool employs the BEM method described in Chapter 11. Figure 15-2 shows the restored pavement at a similar keyhole excavation, where an 18-in. core was extracted and saved, then adhered back into place. Using this method, the road can be returned to service within a couple of hours, and no long-term damage to the pavement occurs.

Source: HDR

Figure 15-1 Electromagnetic scanning of pipe within keyhole excavation

Source: HDR

Figure 15-2 Keyhole excavation with pavement repair

NONDISRUPTIVE SCREENING TECHNIQUES

Screening methods provide relatively low-resolution data, which some utilities use to determine the general condition of a pipe. While these methods cannot find specific defects and thus cannot precisely gauge the risk that a break will occur, they provide information regarding the general extent of deterioration.

Acoustic Velocity Testing

Chapter 10 describes acoustic velocity testing, which is a relatively inexpensive, noninvasive method of determining average pipe condition. Several utilities consider this an effective screening method for determining areas where significant degradation has occurred. This method may be particularly applicable for pipes where the corrosive environment is relatively uniform and well understood. Examples are the insides of unlined cast-iron and asbestos cement pipes or poorly protected metallic pipe in relatively uniform soils. Figure 15-3 shows acoustic velocity testing applied on a Los Angeles trunk line. In addition to providing information on average wall thickness, this method also detects leaks.

Pipe Wall Analysis Using Magnetometry

As a result of a US Environmental Protection Agency–sponsored research project, a company offering in-pipe leak detection tools discovered that magnetometer data collected by one of its in-pipe tools appeared to signal areas where pipe corrosion had occurred. The analysis of these data to find pipe wall anomalies was briefly offered by the company (in addition to the leak detection and gas pocket detection for which the tool is commonly used), but the method has been withdrawn from the market, pending additional study. See Appendix A for more information regarding this method.

Push-in Video and Audio Probe

To determine the presence or absence of mortar lining and make a general assessment of lining conditions, the use of push-in cameras can be helpful. By assessing the interior

Source: HDR

Figure 15-3 Acoustic velocity testing on a Los Angeles trunk line

condition, a general assessment of interior corrosion can be made. These push-in probes (shown in Chapter 7) can also be equipped with hydrophones for leak detection. One of the newer push-in probes is also equipped with an ultrasonic device that reportedly scans the pipe wall in various locations, but no independent reports are available to confirm its effectiveness.

LEVERAGING CONDITION ASSESSMENT DATA

Some mains are more readily assessed than others, due to their location, size, operational conditions, or accessibility. Where mains are difficult to assess, their conditions can sometimes be inferred from that of other mains, provided they are of similar age and material and exposed to a similar environment. By leveraging condition assessment data over multiple mains, the cost of the assessment is effectively reduced.

An example is leveraging the results of high-resolution scanning over several mains. In Chapter 11, an in-pipe tool that employs remote-field electromagnetic testing (RFT) to scan the pipe wall is described. This tool can be inserted at fire hydrants (Figure 15-4) to scan 6-in. and 8-in. mains and has been used in this manner since the late 1990s. However, this method is not as readily deployed on larger pipes as the larger pipes require larger tools that cannot be inserted through fire hydrants. Moreover, the method is relatively disruptive and costly as it involves shutting down the main for several hours. Using this method where it is relatively easy, and extrapolating data to other mains, makes it more cost-effective. This is particularly true if the work is part of a large program where many mains are scanned over several weeks of work.

Source: HDR

Figure 15-4 Inserting an RFT tool into the hydrant lateral of a shallow-buried main

The City of Calgary adopted this tool and has been scanning mains for 15 years, using the results to determine which pipes should be replaced and which pipes to protect with galvanic anodes. Calgary credits this tool with helping reduce the break rate, even as the main replacement rate was cut. The accuracy of this tool has been validated in several independent studies, including the recent Water Research Foundation Project 4471, "Leveraging Data from Non-Destructive Examinations to Help Select Ferrous Water Mains for Renewal" (Ellison et al. 2017).

A riser pipe extension has been added to make it easier to insert the RFT tool into the vertical pipe. The blue "hydrant" is provided by the testing company. Clamps are used to hold it in place, rather than trying to match bolt patterns. The tool will be pulled through the pipe by the foam pig (swab) that is being inserted first.

The deployment of this tool through hydrants is relatively low risk and can be cost-effective (as demonstrated in Calgary). The notion of inferring the condition of non-inspected mains using data from similar mains applies to other inspection techniques, provided that the mains are very similar in terms of age, material, and environmental exposure and the data are transferable. The more such methods are employed, the better our knowledge will be regarding data transference.

THE ASSESS-AND-FIX APPROACH

The assess-and-fix concept is to perform condition assessment at the same time that a pipe rehabilitation project is underway. The decision to rehabilitate a pipe may be made using various factors, as is currently practiced, but the method of rehabilitation is not selected until the pipe is scanned and its condition is determined. Unlined cast-iron pipes are classic candidates for an assess-and-fix approach, but so is any pipe whose condition is believed to be compromised and where a trenchless method of renewal could be beneficial.

The difficulties and risks associated with scanning a water main are essentially eliminated if the inspection is performed as part of a cleaning and lining rehabilitation project. If scanning is performed after the cleaning but before the lining, the added field effort is minor. The scanning tool can be pulled through the pipe at the same time that a final video inspection is often performed. The NDE data can then be evaluated and a lining selected and designed to remedy integrity issues with the host pipe. On a project involving multiple mains, crews could be directed to other work while the engineering evaluation is completed without significant delay to the project. An assessment during rehabilitation is thus very manageable. This method has been demonstrated in both Phoenix and in Washington, D.C., where the value of the assessments was also demonstrated.

The steps involved is assess-and-fix lining are as follows.

1. Select a main for rehabilitation/renewal
2. Establish a contract suitable for the method
3. Set up a bypass piping system
4. Excavate and remove a pipe segment for access to the pipe interior
5. Clean the main
6. Assess the main
7. Select and design the lining
8. Apply and cure the lining
9. Close up and test the pipe (pressure and bacteria)
10. Reinstate the services
11. Return to service
12. Complete the project

Of these steps, only numbers 6 and 7 are added; all the others are part of traditional water main rehabilitation. Figure 15-5 shows insertion of an assessment tool in the City of Phoenix as part of Water Research Foundation Project 4473, "The Assess-and-Fix Approach: Using Non-Destructive Evaluations to Help Select Pipe Renewal Methods" (Ellison et al. 2017). This free, downloadable (www.waterRF.org) report provides all the information needed for utilities to implement an assess-and-fix method.

FORENSIC ANALYSIS OF BREAKS

When a water main breaks prematurely and without apparent cause, forensic analysis may be appropriate. Through forensic analysis, breaks have been traced to materials that did not meet specification, construction that was poorly performed, and operational events (surges) that were not considered during design of the pipelines. By finding these causes, mitigations can be implemented to avoid future breaks.

The word *forensic* refers to "debate" and implies that the investigation may be part of a legal dispute. For this reason, it is often important to perform the investigation in a very deliberate manner, exploring all possible causes and avoiding premature conclusions. Each step should be well documented, and every communication (e-mails in particular) should be carefully worded to avoid any implication that the investigation is biased against any particular party. While few utilities embrace litigation, considerable money can be recovered from contractors and material suppliers (often through negotiations) by conducting thorough, scientific investigations of failures. The impending failures of PVC and other pipe materials in particular are nearly impossible to detect, which makes it very important to find the causes of failures so that appropriate actions can be taken. Figure 15-6 comes from a laboratory report where impingement on a rock was found to be the ultimate cause of a PVC pipe break.

Source: Water Research Foundation, Report 4473.

Figure 15-5 Photo showing insertion of the RFT tool during assess-and-fix demonstration in Phoenix

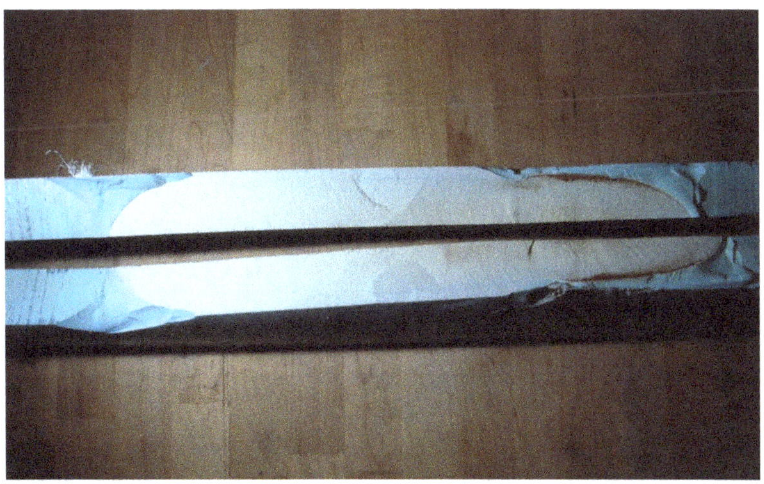

Source: HDR

Figure 15-6 Fracture analysis of a PVC pipe break

REFERENCE

Ellison, D.; Spencer, D.; and Brander, R. 2017. Leveraging Data from Non-Destructive Examinations to Help Select Ferrous Water Mains for Renewal. Denver: Water Research Foundation.

Chapter 16

The Next Steps: Using Condition Assessment Information

Nathan Faber, Chapter Lead, San Diego County Water Authority
Jeff Leighton, Portland Water Bureau (retired)

After condition assessment information has been gathered, it can be used to make decisions and help manage infrastructure. This chapter briefly addresses the purposes of condition assessment, the management of data, the analysis of those data, and the presentation of results. Each of these items is critical for the success of a condition assessment program.

PURPOSE OF CONDITION ASSESSMENT INFORMATION

The data obtained from pipeline condition assessments helps drive decisions on pipeline repair or renewal based on risk, fiscal impact, and utility policies. The purpose of a condition assessment effort is to obtain data to support pipeline management decisions. In most cases, the utility makes pipeline management decisions as part of a risk-based strategy, as discussed in Chapter 1.

There are several pipeline management decisions that can be grouped into three main categories: immediate action, long-term action, or no action. The utility can execute immediate action during the condition assessment effort while the pipeline is out of service or shortly after the condition assessment project. In many cases, these actions are short-term fixes until the utility can execute a long-term action. There are several options for long-term actions, including pipe replacement and trenchless rehabilitation. Trenchless rehabilitation options are described in AWWA M28, *Rehabilitation of Water Mains* (2014). The utility can develop a plan for overall pipe renewal strategies for their particular needs. Strategies can include, for instance, areas of targeted repairs, larger areas of long pipeline reaches that may need rehabilitation, or a combination of both approaches. These strategies can be integrated into the utility's overall capital improvement program. As part of a no-action decision, the utility may decide to continue to monitor the pipeline condition with future condition assessment efforts. In addition, the utility can execute a remaining pipeline life analysis to estimate a time frame for future action. Pipeline management decisions should account for the detail level of the condition assessment technology applied and the specific pipe material.

The utility can also use the condition assessment information to derive follow-up actions to extend pipeline life. For example, a utility may determine a future assessment schedule to capture decay rate or take proactive steps to monitor the pipe condition decay based on review of the results.

LEVERAGING DATA

Risk-based decision making requires that data from several different sources be analyzed and compared, including condition data and pipeline asset data. Proper management of data can help identify spatial information for consequence of failure using a geographic information system (GIS) and correlate several different data sets such as condition, soils, maintenance, design, construction, operation, and other data.

A data management strategy, such as a central database, is valuable because it allows for data to be stored in a central repository for wider access, cross-referencing of various data sets, single-source updates, and better utilization across utility groups because the data are consistent and accessible. It also helps for long-term rehabilitation strategies to accurately identify projects and combine multiple efforts into a single project.

It is important to integrate various data sets into a usable format. For pipeline assets, a single unique identifier for the utility's distinctive pipeline assets is important. For example, pipeline station may be repeated for several different pipelines in the system; therefore, data with pipeline station may not apply to a single asset and require a unique identification. A unique identification should be included in all condition assessment information.

Also, it is important to capture important condition data throughout the life of the pipeline, including leaks, repairs, construction defects, as-built conditions, operational changes, and manufacturing information. Various information systems can be used to manage condition data. For example, a GIS or other database can track the physical location and asset specification data. A computerized maintenance management system (CMMS) tracks events and changes at a more transactional level, such as breaks, condition, work, and costs. Mobile devices provide direct field capture for important information that integrates back to the GIS, CMMS, or other databases.

There can be different resolutions and levels of detail for breaking down pipeline data, such as pipeline contracts, valve-to-valve reaches, street limits, material type, diameter, and/or pipe sections (or pipe sticks). Contract data can consist of the same types of pipeline materials, built under a single contract and a single pipe supplier. The data and condition can generally follow the same trends. Various condition assessment methods provide different data resolutions, from pipeline reaches down to pipe section level. For example, acoustic testing may provide information for a pipeline from valve to valve. Ultrasonic testing can provide specific information at a specific location on a pipe section with high resolution. It is important for the utility to understand the resolution of the condition data the project will provided before it starts. It is also important to understand how the utility will integrate the data into its existing data structure. This will help efficiently integrate the condition data with existing data and provide the ability to use the data efficiently to make decisions.

Utilities need accurate information on pipeline assets to get the most benefit from condition assessment efforts. For some utilities with limited information on their pipelines, this process can be overwhelming and can limit some of the condition assessment technologies that can be used. An effective strategy could be a stepped approach starting with basic data such as material type, diameter, and installation date. Over time, the utility can build on the base data and add information such as leaks and pipeline depth moving forward with a sound data structure and quality.

DATA ANALYSIS

There are various levels of condition assessment data analysis, from simple review of the data to a detailed engineering structural evaluation to remaining pipe life estimations. Structural evaluations depend on pipe materials. Finite element modeling for detailed analysis of most pipe materials can provide information on the structural capacity of a pipe compared to damage or decay information. For prestressed concrete cylinder pipe, finite element methods can provide risk curves to compare potential pipe failure with wire break distress. For metallic pipelines, there are currently two main resources for analysis ASME B31G (2012) and API 579-1/ASME FFS-1 (2007). These resources can help evaluate the remaining strength of corroded pipelines based on general metal loss, local metal loss, pitting corrosion, cracks, dents, gouges, and laminations.

Utilities can estimate remaining life in several different ways based on different factors. Remaining life can be estimated based on an analysis of historical pipe failure data and condition factors. The factors can be weighted and calibrated with historical data. Then, these factors are projected to an existing pipeline to estimate the pipeline's remaining life. In addition, condition rate of change information, measured with a monitoring system or multiple condition assessments, can be used to estimate a time of pipe failure.

Utilities should understand and evaluate through condition assessment all potential pipeline failure modes to fully determine the overall condition of the pipe. For example, a condition assessment technique may provide specific data on leaks; however, it might not provide information on other important pipeline failure modes such as corrosion or fatigue. This is different than the confidence in whether a particular condition assessment technique can detect the flaws for which it was intended.

PRESENTATION OF RESULTS

Results of condition assessments can be presented in several ways, including spreadsheets, tables, graphs, matrixes, and GIS and three-dimensional graphics. The selection of these may be driven by the intended audience, desired messages, and/or type of information conveyed.

Typically, condition assessment projects will have a written report covering information regarding the project details and technology calibration, accuracies, and limitations. The report may also have helpful summaries of the detailed data including maps, graphs, and charts with conclusions on the overall results. The report generally contains detailed results presented in a table format and an associated electronic spreadsheet and/or a GIS file for results that are used by the utility.

Data in a table can be divided into various resolutions and include various attributes that can be associated with each pipe, such as stationing, unique pipe identification, pipe section number, length, and condition information. These data can be summarized with

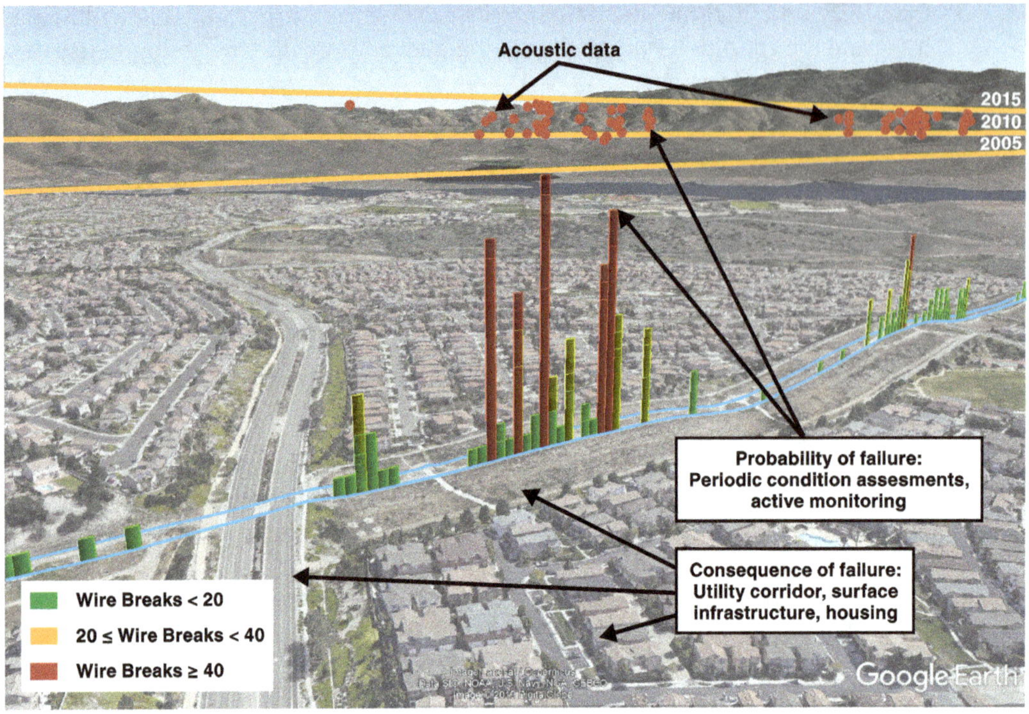

Adapted with permission from San Diego County Water Authority. Image Landsat / Copernicus; Data SIO, NOAA, U.S. Navy, NGA, GEBCO; Image © 2019 DigitalGlobe.

Figure 16-1 Condition assessment data displayed in Google Earth

graphs or charts to identify overall trends for the system or pipeline, such as number of pipes with decay and locations of decay. If risk is identified, this can also be graphed with probability of failure and consequence of failure in a graph or matrix. Graphics can also be used to provide profiles showing individual pipe sections or two- and three-dimensional pipe images with data.

GIS can help evaluate the consequences of failure, including potential impacts to surrounding area, by spatially displaying the pipeline in relation to other assets (e.g., roads). The data can also interface with other types of data already available in the utility's GIS system, such as soil data, right-of-way ownership, drainage structures, facilities, and district boundaries. Data from the GIS system can also be displayed in three dimensions and/or exported into accessible software such as Google Earth for display (Figure 16-1).

The information can be displayed with spatial software in the context of elevation changes on the pipeline, imagery changing over time, and condition data that can be displayed in three dimensions and change over time (Coghill and Yek 2016).

EXAMPLE APPROACHES

WaterOne of Johnson County, Kans., performed a condition assessment of its 30-in. unwrapped, ductile-iron transmission main using electromagnetic remote field technology (RFT). Based on the condition assessment data and report, they were able to quickly evaluate the RFT data using a bar graph representing the overall condition of the pipeline. The graph showed the three worst corrosion pits for each pipe segment in that section. Then, a diagram showed them the overall locations of the corrosion spots in relation to both longitudinal and clock position. Finally, a complete wall thickness summary table

Circumferential Wall Thickness		Local Wall Thickness *Clock positions are with west to east perspective (ie 3:00=south, 9:00=north). Local wall loss entries in red are pitting indications with ≤15% remaining wall.								
		Tmin1			Tmin2			Tmin3		
Tcircmax RW (%)	Tcircmin RW (%)	RW (%)	Location (ft)	Clock Position	RW (%)	Location (ft)	Clock Position	RW (%)	Location (ft)	Clock Position
102%	95%	53%	20.16	1:00	57%	9.51	0:30	76%	11.70	0:30
105%	96%	24%	28.77	1:00	28%	28.22	1:00	80%	40.04	7:00
100%	94%	36%	60.13	0:30	42%	59.38	1:00			
102%	96%	44%	75.42	5:00	52%	67.52	1:00	70%	79.89	1:00
98%	92%									
104%	98%	72%	106.98	5:00	79%	110.67	6:00	80%	110.10	6:00
99%	92%	60%	127.21	0:30	68%	130.13	6:30			
98%	93%	71%	149.08	4:30						
99%	91%	17%	165.48	1:30	45%	165.57	12:00	59%	177.62	0:30
103%	95%	27%	197.71	0:30	58%	189.36	6:00			
102%	95%	16%	211.73	7:00	41%	209.02	6:30	48%	217.65	0:30
97%	91%	22%	242.37	1:00	70%	237.56	5:00	76%	236.38	5:00
103%	98%	4%	261.62	0:30	11%	260.23	0:30	20%	256.29	1:00

Adapted with permission from WaterOne

Figure 16-2 Summary table

(Figure 16-2) provided them with the detailed information for the three worst wall loss locations in relation to both longitudinal and clock position on each pipe segment.

WaterOne used these detailed data to develop a risk analysis of the pipeline to determine a prioritized, long-term, staggered replacement schedule; sections of immediate repairs with clamps and/or short sections of pipe replacement including cathodic sacrificial anodes to extend the life of the pipeline; and areas to monitor for leaks and cathodic sacrificial anodes. This approach delayed a full pipeline replacement, leading to significant savings. (Pietig 2015)

In 2004, City of Dallas Water Utilities (DWU) implemented a structured leak detection program to address nonrevenue water (NRW) within its nearly 5,000-mile potable water system. Over the past decade, the program's focus and worth have significantly changed with the allocation of additional resources and use of improved technologies, including data management techniques. In addition to leak location, technologies used within DWU's leak-assessment program can identify and record air pockets, visual internal pipe wall condition, and in some instances, the average pipe wall thickness. Data obtained from the program are recorded in two separate databases (Large-Diameter Pipeline Assessments and Distribution Pipeline Replacement Management Database), where a series of weighted scores are applied for a suitable reliability and risk rating. The final ratings are used in the decision-making process for corrosion protection systems, pipeline reinvestment projects and operational decisions. Particularly for larger-diameter PCCP pipelines, 48-in. through 108-in. in diameter, the results and data acquired are often used as a precursor for other structural assessments such as electromagnetics. DWU's systematic approach and data management techniques have enabled it to effectively manage its capital reinvestment program, lowering its NRW from 11% to 7% and improving its system integrity rating from 0.6 to 0.3. (Payton 2015).

The San Diego County Water Authority (Water Authority) has been using advanced condition assessment methods for its PCCP since 1999 and for its steel pipelines since 2011. Over the years, it has used condition assessment data in several ways, starting with a calculation of an estimated useful life to a risk-based, three-dimensional graphical evaluation. In 1997, the Water Authority developed estimated service lives for some of its pipelines using several weighted factors including design, construction, operations, maintenance, and environment. Initially, these estimates included visual and sounding condition assessments and later included data from remote field eddy current condition assessment. In addition to identifying areas for immediate repair, these efforts led to a repair schedule for PCCP. With the implementation of PCCP monitoring in 2006 and the use of magnetic flux leakage for steel pipelines in 2011, the Water Authority transitioned to a centralized, open source database (Faber and Hill 2016) and three-dimensional data display to develop a risk-based rehabilitation schedule (Coghill and Yek 2016). This has led to the transition to a targeted rehabilitation approach and the deferral of a significant amount of pipeline rehabilitation costs.

REFERENCES

API (American Petroleum Institute). 2007. API 579-1/ASME FFS-1. 2007 (2nd ed.). *Fitness-For-Service*. Washington: American Petroleum Institute.

ASME (American Society of Mechanical Engineers). 2012. B31G. *Manual for Determining the Remaining Strength of Corroded Pipelines*. New York: ASME.

AWWA (American Water Works Association). 2014 (3rd ed.). M28, *Rehabilitation of Water Mains*. Denver: AWWA.

Coghill, M.R. and Yek, C. 2016. Visualizing Buried Risks: The Interpretation of Condition Assessment Data for San Diego County Water Authority's Large Diameter Pipelines. American Society of Civil Engineers. Proc. 2016 Pipelines Conference, Kansas City, Mo.

Faber, N.D. and Hill, R.C. 2016. Pipeline and Survey Data for Long-Term Asset Management. American Society of Civil Engineers. Proc. 2016 Pipelines Conference, Kansas City, Mo.

Payton, R.L. 2015. Proactive Pipeline Assessment Saves Millions of Gallons of Water and Prevents Catastrophic Pipeline Failures. Proc. 2015 CA-NV Section AWWA Fall Conference, Las Vegas.

Pietig, S. 2015. The Condition Assessment of a 30-inch Ductile Iron Water Line by WaterOne of Johnson County, Kansas. Proc. 2015 Pipelines Conference, Baltimore.

Appendix A

Other Assessment Methods

The methods briefly described here have not become well-established in the North American market as of completion of this manual. For additional information about these methods, please contact the technology companies.

ACOUSTIC RESONANCE TESTING

A Norwegian company adapted a method originally developed to assess the condition of a sunken ship for inspection of iron and steel water mains using an in-pipe tool. A transducer transmits an acoustic signal toward the pipe wall. The signal propagates in the pipe wall, exciting resonances, which are detected by a receiving transducer. Analysis of the response determines the resonance frequencies, from which the thickness of the pipe wall can be estimated (https://breivoll.eu/diagnostics/art-technology).

PIPE WALL ANALYSIS USING MAGNETOMETRY

A company offering in-pipe leak detection tools discovered that magnetometer data collected by one of its tools appeared to signal areas where pipe corrosion had occurred (Figure A-1). For a brief period, the company offered analysis of these data to detect pipe wall anomalies, but this service is not currently marketed pending further research.

The tool records changes to the magnetic susceptibility of a material that are created by variations in mechanical stress. This long-recognized phenomenon is technically described as the "inverse magnetostrictive effect" (also called "the Villari Effect"). The variations in mechanical stress may be caused by thinning of the pipe wall or by other factors such as inclusions or other defects in the pipe material, bending of the pipe, and

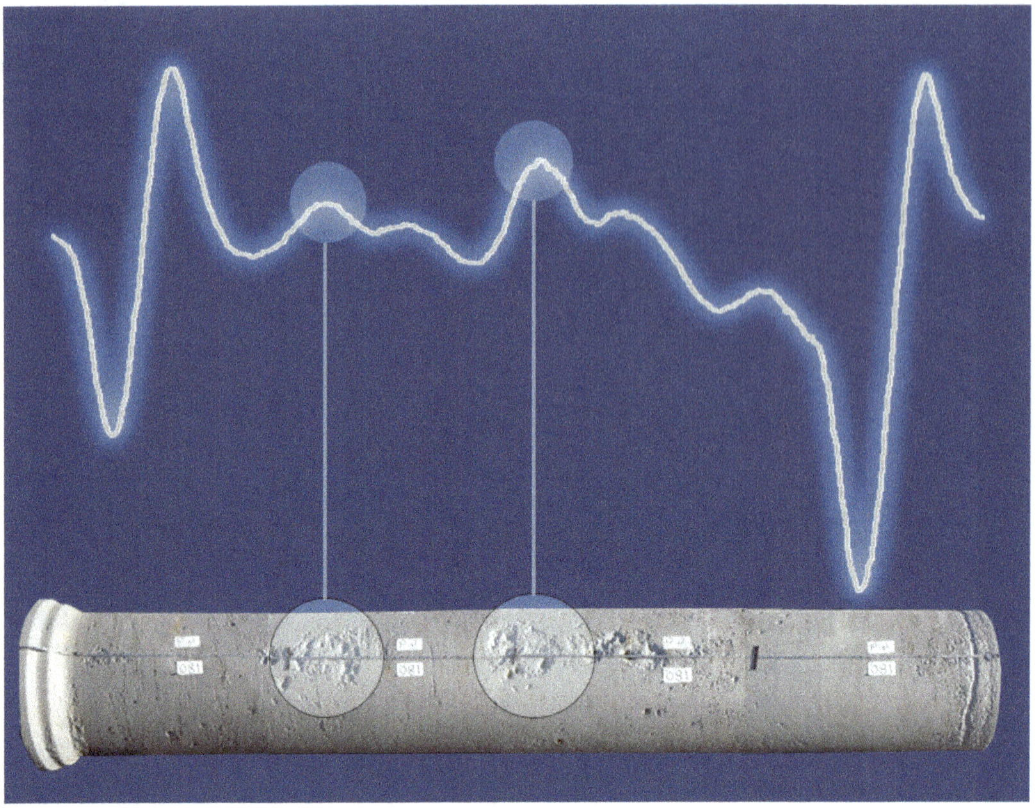

Source: Pure Technologies
The analysis of magnetometer data collected by an in-pipe leak detection tool seemed to show signals that corresponded to corrosion on this 24-in. cast-iron main.

Figure A-1 Pipe wall analysis using magnetometry

rock impingements. The tool is very compact and relatively easy to deploy and can often be used with the pipeline in service.

Additional research is needed to determine how accurate and precise the inspection results are. Excavations performed where anomalous readings were detected have often found corroded pipe or other causes of elevated stress. A side-by-side comparison of this testing and electromagnetic scanning, however, was less positive. Because the detection tool rolls along the bottom of the pipe, its ability to detect small defects in the pipe crown will be limited, particularly if the pipe is large diameter.

Index

Note: *f.* indicates figure; *t.* indicates table

A

absolute risk score, 3
accelerated corrosion testing, 72
accessibility, pipeline, 21
acoustic leak detection, 79–91
  external acoustic methods, 80–87
  internal acoustic methods, 88–91
  PCCP, 178
acoustic resonance testing, 207
acoustic testing velocity, 125–133
  background, 127–130
  history, 126–127, 126f
  low-value pipes, 195, 196f
  performance, 130–132
  requirements and limitations, 132–133
acoustic velocity techniques, 58
acoustic wire break monitoring, 177–178
adhesion testing, 63
advanced sensor technology, 106–110
aging curve, 43
air pockets, 133
Anchorage Water Wastewater Utility, 10, 11f
anomaly, 120
appurtenances, 133
asbestos cement pipe, 131
  stain test, 73, 73f
"assess-and-address" approach, 10
assess-and-fix approach, 6–7, 197–198, 199f
asset, 2
asset-management strategies, for pipes, 2–3
aural testing of concrete, 65

B

Barnes Layer method, 50
break avoidance, 3–4
break rates, 34, 36
  PCCP manufactured in different areas, 167f
breaks, 34, 36–37
  forensic analysis, 198–199
  prevention, 78–79
broadband electromagnetics (BEM), 136, 140, 145
budget, for MFL inspection, 162
business cases, for condition assessment, 10–12

C

carbonation, 71
cast iron pipe, 131
  depth of corrosion pits in, 62f
chemical analysis, of pipe materials, coating, soil or water, 68
chloride, 63, 68
circumferential position reference, 121
City of Calgary, Alberta, 15
close-interval survey (CIS), 51, 52–53, 52f, 54f
closed-circuit television (CCTV), 23, 88
  for dewatered pipes, 102
  images from inspection of a 21-in. steel water main, 98f
  internal visual inspection, 101–105
  tethered for fully operational/pressurized pipes, 102, 105, 105t
coating assessment, 63
coating inspection, visual inspection and, 61
coatings, 132
communication, 27–29
communications network, 85f
concrete-composite, 132
concrete/mortar inspection, visual inspection and, 61
condition assessment
  building support, 9–16
    accounting treatment, 12–13
    budget for, 13–14
    business case for, 9–12
    policy makers, public and other stakeholders, 14–16
  internal remote visual inspection, 98
  next steps, 201–206
    data analysis, 203
    example approaches, 204–206

leveraging data, 202–203
presentation of results, 203–204
purpose of information, 201–202
planning for, 19–30
progressive, and resulting actions, 14
project/program planning, 19–30
available technologies, 23–27, 24t–26t
choosing a method, 21–23
contract guidelines, 29–30
levels, 19–21, 20f
planning of field work/communication, 27–29
using sonar, 106–108
for water mains, 1–8
additional considerations for, 7–8
benefits and costs of, 3–6, 5t
definition of, 2–3
as part of an assess-and-fix strategy, 6–7
consequence of failure (CoF), 2–3
desktop scoring, 39–40, 39t, 40t
contingency planning, 161
continuity testing, 51
controlled destructive examination, 186
correlators, 81–82
comparison of leak detection, 82t
parameters, 82t
corrosion, 47
in MFL data, 157f
PCCP damage, 166
corrosion potential survey, 51
corrosivity surveys, 178
cost-benefit analysis, 4–6, 5t, 6t
cost-effectiveness, 13
costs
hydrostatic examination, 187
inspection methods and technologies, 112, 112t
cracking, PCCP inspection, 167, 169f
Dallas Water Utilities (DWU), 205
data cleanup, 33–34
data collection, 45
data handling, 161
data sources, 32–33
decay curve, 43, 43f
defect, 120, 154

D

Denver Water, 37
deployment platform, 171
desktop condition assessment, 2, 31–45
data cleanup and assessment, 32–36
likelihood of failure evaluation, 37–42
long-term renewal planning, 43–45
PCCP, 166, 167f
setting level of service goals, 36–37
digital ultrasonic gauge, 63
direct assessments
PCCP, 166–175
distance measurements, 133
distribution mains, 32
Distribution System Investment Charge (DSIC) program, 12
distribution system monitoring, 84–86, 85f
documentation, 121
"double block and bleed" method, 118
"drive-by" data collection, 83
dry film thickness (DFT), 63
ductile iron, 131
effective useful life (EUL), 43–44, 44f
electrical continuity testing, 51, 64
electrical currents, 170, 171f
electrochemical impedance spectroscopy (EIS), 66
electromagnetic conductivity survey (Emag), 49–50, 50f
electromagnetic inspection, PCCP, 170–175, 172f
electromagnetic testing technologies, 135–147
advantages and limitations, 146–147
available tools, 141t–144t
defect resolution considerations, 146
free-swimming tool, 137f
handheld tool, 138f
high-resolution, large-diameter tool, 137f
history and current use in water mains, 136–139
hydrant-launched, water-propelled tool, 139f
manually winched tool, 139f
self-propelled crawler, 138f
state of the practice and application, 145–146
technology overview, 140–146
elemental analysis, 68–69
eletrochemical testing, 65–66, 66f
embedded cylinder pipe (ECP), 163–164, 164f
EM inspections results for, 174
embrittlement, 165
energy-dispersive spectrometry (EDS), 68–69
identification of corrosion products, 69f

excavation
  precautions, 60
  utility clearance and permit, 59–60
external corrosion direct assessment (ECDA), 47–55
  direct examination, 53–54
  economic benefits, 55
  indirect inspection, 49–53
  post-assessment, 54
  pre-assessment, 48–49

E

external direct assessment, 20f
external inspection
  EM inspection access requirements, 173, 174f
  PCCP, 170

F

failure analysis, 22–23
  leak detection and, 78
  using risk curves, 181
failure modeling, 38–39
feature, 120
fiber optic sensor cable, 177, 178f
field condition assessment, 2
field work, planning of, 27–29
fluids, 133
forensic analysis, of breaks, 198–199
forward-looking sonar, 106, 107f
free-swimming devices, 89–91, 90f, 91t
free-swimming multisensor equipment, 105

G

galvanic cathodic protection system, 179
gas tracer method, 91–92, 91f, 92t
geographic information system (GIS), 35
  based CoF scoring criteria, 39, 40f, 40t
  condition assessment data, 204f
  prediction of funding needs based on service levels, 45f
ground-penetrating radar, 92, 93t

H

health considerations, 28–29
helium, 91
hydrostatic testing, 185–191, 187f
  after initial tests, 191
  application, 188, 188t
  basic, 186
  pressure limits, 190–191, 190t
  procedure development, 188–189

I

impressed-current cathodic protection system, 179
in-line inspection (ILI) tool
  MFL technology, 153–155
    CML pipeline, 154f
    examples of inspection, 155–157, 156f
    test data in pipe with mortar lining, 155f
in-pipe acoustic tools, 130, 130f
in-pipe condition assessment, 20f
in-pipe leak detection tools, 23, 189
in situ field testing, 60–67, 60t
  aural testing of concrete, 65
  microbiologically influenced testing, 66–67
  rebound testing, 65
  in situ chloride content, 63
  in situ coating assessment, 63
  in situ electrical continuity testing, 64
  in situ elemental analysis of metal pipe using x-ray fluorescence, 66
  in situ eletrochemical testing, 65–66, 66f
  in situ pH testing, 61–63
  in situ potential mapping, 64
  visual inspection, 61
  wall thickness and pit-depth measurements, 63–64
indirect assessments, 175–178
inspection tools, 155
interference, 133
  stray-current, 179
internal corrosion direct assessment (ICDA), 55
internal inspections, 28
  EM inspection access requirements, 173
  PCCP, 166–167
internal remote visual inspection, 97–113
  condition assessment tool, 98
  current available methods and technologies, 101–110, 102t
  implementation strategies, 110–112
  logistical considerations, 100–101
  recent advances in technology, 99–100, 99f
inverted condition assessment pyramid, 13–14, 14f

K

keyhold methods, 194, 195f

L

laboratory testing, 68–73
lamination, 154
laser scan, for pipe wall surfaces, 103f
lasers
  above-water, 109, 110f
  condition assessment using, 108–110
  underwater, 108–109, 110f
leak, 34, 36–37
  prioritizing pipes for inspection, 78
leak correlator method, 128–129
  typical setup, 128f
leak detection, 77–96, 80f
  acoustic, 79–91
  data for condition assessment, 78–79
  how and when to perform, 79
  monitor in below-ground chamber, 84f
  node in hydrant cap, 85f
  non-acoustic, 91–96
  sensor connected to a service line, 85f
leak detection methods, 23
leak loggers, 83, 83t
leak noise correlation, 23, 82
Leakage Component Analysis (LCA), 79
level of service (LOS), 36
likelihood of the failure (LoF), 2–3
  evaluating desktop score, 37–42
linear polarization resistance (LPR), 66
lined cylinder pipe (LCP), 163–164, 164f
linings, 132
listeners, 81, 81t
lockout-tagout (LOTO) procedures, 118
longitudinal position reference, 121
Los Angeles Department of Water and Power (LADWP), 186, 189f
low pressure, 133
low-value pipes
  economical assessments, 193–199
    assess-and-fix approach, 197–198
    forensic analysis of breaks, 198–199
    leveraging condition assessment data, 196–197
    nondisruptive sampling methods, 194
    nondisruptive screening techniques, 195–196
low-voltage conductivity, 93–94, 94f, 95t

M

magnetic flux leakage (MFL), 4–5, 149–162
  application considerations and inspection logistics, 150–157
    advanced principles, 150–153, 151f, 152f
    fundamental principles, 150
  current state, 150
  data output and post processing, 157–160
    C-scan of blind test defects, 158f
    performance specifications, 159–160, 160t
  integrity management, 162
  planning and delivering a quality project, 160–162
magnetometry, in pipe wall analysis, 207–208, 208f
metallic pipe inspection, visual inspection and, 61, 62f
metallographic analysis, 70, 71f
microbiologically influenced corrosion (MIC), 66–67
  sample slime for testing, 67f
microscopic analysis, 69, 70f
monetized approach, for CoF, 39

N

node, 86
non-acoustic leak detection, 91–96
  gas tracer method, 91–92, 91f
  ground-penetrating radar, 92, 93t
  low-voltage conductivity, 93–94, 94f, 95t
  satellite-based leak detection, 94–96, 95f, 96t
  thermography, 92–93, 93t
non-invasive methods, 20f
nondestructive examination (NDE), 187

O

opportunity assessments, 22–23, 59
over-the-line surveys, 58

P

petrographic examination, 70–71, 71f
pH testing, 61–63, 62f
phenolphthalein, 61, 63
  stain testing, 72, 73f
photos, visual inspection and, 61
physical entry inspections, 115–124
  electronic data collection, 122f

interpretation and presentation of results, 122, 124
pipeline entry limitations, 116, 117f
planning, 116–122
safety considerations, 117–119
pipe bends, 101
pipe coupon
AV tests, 127
EDS on, 68–69
pipe length, 101
pipe overload, 166
pipe slope, 101
pipe-to-soil potential survey *see* corrosion potential survey
pipe wall analysis, 207–208, 208f
pipeline
access points to, 100–101
anomalies, 158t, 159f
B-H curve, 152f
components, 133
constraints due to configuration, 101
entry limitations, 116, 117f
isolated defects, 156f
operating conditions, 101
pipeline access, 161
pipeline geometry, 161
Pipeline Operators Forum, 157
pipes
defined, 35
external system access requirements, 132–133
hydrostatic testing, 185–191
PVC, fracture analysis of break, 199f
statistical analysis, 35–36
visual inspection and, 61
pit-depth, 63–64
polarization tests, 72
potential mapping, 64, 65f
predictive failure analysis, 7–8
pressure monitoring, 176
pressure wave analysis, 129
prestressed concrete cylinder pipe (PCCP), condition assessment, 163–182
construction, 163–165
destressed, strategies for management of, 178–182
evaluation, 181–182
inspection, 180
monitoring, 180–181
protection, 179
rehabilitation and replacements, 182
deterioration and failure modes, 165–166
inspection and assessment, 166–178
probability of detection (POD), 158
profiling sonar, 106, 107f
project team, 161
pulsed eddy current (PEC), 136

R

reasonably certain, 12
rebound testing, 65
redundancy, 40–41
relative risk score, 3
remote-field eddy current (RFEC), 136
remote field technology (RFT), 5–6, 140–145
probe, 140f
in shallow-buried main, 197f
remote systems, 83–84
remotely operated vehicles (ROVs), 99
with CCTVs, 103f
condition assessment using, 108, 108f, 109f
high-definition camera in dry conditions, 104f
high-definition camera with zoom capability, 104f
swimming, 105
renewal planning, 4
repair segments, 133
repairs, water mains, 22
replacement costs, 44
request for proposals, 29
risk analysis, 2–3
risk mapping, 42f
risk matrix, 41f
risk scoring, 40–41
robotic inspection platforms, 169, 171, 172f

S

sacrificial anode stray-current drains, 179
safety considerations, 28–29
salt spray test, 72, 72f
sample locations, 194
San Diego County Water Authority, 206
satellite-based leak detection, 94–96, 95f, 96t
scanning electron microscopy (SEM), 69
Schmidt Hammer test *see* rebound testing
self-contained breathing apparatus (SCBA), 118
service goals, 36–37
long-term funding needs and, 44–45

service interruptions, 23
service outage, 161
service taps, 22
serviceability limit, 181
simulative analysis, 72
soil resistivity, 49–50
soil sample, 51
soils, visual inspection and, 61
sonar
  in condition assessment, 106–108
  limitations, 108
sonic and ultrasonic technology, 175, 176f
sounding inspection, 166–170, 168f
sounding survey, 80, 81f
spot assessment, 57–74
  leak detection and, 78
  post assessment and interpretation of results, 73–74
  strategies, 58–60
  testing procedures, 60–73
statistical analysis, of pipes, 35–36
statistical studies, 20f
steel, 131–132
stray-current interference, 179
submersible crawlers, 105
sulfate, 63

T

tethered leak detection devices, 88–89, 89f, 89t, 100f
  CCTV, 105t, 106f
thermography, 92–93, 93t
transient monitoring and mitigation, 176–177
transmission electron microscopy (TEM), 69
transmission main monitoring, 86–87, 86t, 87f, 87t
transmission mains, 32
"triple bottom line" analysis approach, 39

U

ultrasonic thickness measurement, 64f
underwater inspections, 119
Utility Benchmarking survey, 36
utility clearance, 59–60

V

validation, 161–162
validation database, 127f
visual inspection
  PCCP, 166–170, 168f
  physical entry inspections, 120
  physical testing and, 122, 123t

W

wall thickness, 63–64
  accoustic velocity testing, 129
water mains
  condition assessment, 1–8
  other methods, 207–208
    Water Research Foundation Project 4473, 6
WaterOne of Johnson County, Kans., 204–205, 205f
Wave Velocity Thickness Model, 128
weighted-criteria scoring, 37–38, 38t
  CoF, 39t
Wenner 4-pin survey, 49–50, 50f
wire breaks, 174
  limit estimation, 181

X

X-ray diffraction (XRD), 69
X-ray fluorescence (XRF), 66

AWWA Manuals

M1, *Principles of Water Rates, Fees, and Charges*, #30001
M2, *Instrumentation and Control*, #30002
M3, *Safety Management for Water Utilities*, #30003
M4, *Water Fluoridation Principles and Practices*, #30004
M5, *Water Utility Management*, #30005
M6, *Water Meters—Selection, Installation, Testing, and Maintenance*, #30006
M7, *Problem Organisms in Water: Identification and Treatment*, #30007
M9, *Concrete Pressure Pipe*, #30009
M11, *Steel Pipe—A Guide for Design and Installation*, #30011
M12, *Simplified Procedures for Water Examination*, #30012
M14, *Backflow Prevention and Cross-Connection Control: Recommended Practices*, #30014
M17, *Fire Hydrants: Installation, Field Testing, and Maintenance*, #30017
M19, *Emergency Planning for Water and Wastewater Utilities*, #30019
M20, *Water Chlorination/Chloramination Practices and Principles*, #30020
M21, *Groundwater*, #30021
M22, *Sizing Water Service Lines and Meters*, #30022
M23, *PVC Pipe—Design and Installation*, #30023
M24, *Planning for the Distribution of Reclaimed Water*, #30024
M25, *Flexible-Membrane Covers and Linings for Potable-Water Reservoirs*, #30025
M27, *External Corrosion Control for Infrastructure Sustainability*, #30027
M28, *Rehabilitation of Water Mains*, #30028
M29, *Water Utility Capital Financing*, #30029
M30, *Precoat Filtration*, #30030
M31, *Distribution System Requirements for Fire Protection*, #30031
M32, *Computer Modeling of Water Distribution Systems*, #30032
M33, *Flowmeters in Water Supply*, #30033
M36, *Water Audits and Loss Control Programs*, #30036
M37, *Operational Control of Coagulation and Filtration Processes*, #30037
M38, *Electrodialysis and Electrodialysis Reversal*, #30038
M41, *Ductile-Iron Pipe and Fittings*, #30041
M42, *Steel Water-Storage Tanks*, #30042
M44, *Distribution Valves: Selection, Installation, Field Testing, and Maintenance*, #30044
M45, *Fiberglass Pipe Design*, #30045
M46, *Reverse Osmosis and Nanofiltration*, #30046
M47, *Capital Project Delivery*, #30047
M48, *Waterborne Pathogens*, #30048
M49, *Quarter-Turn Valves: Head Loss, Torque, and Cavitation Analysis*, #30049
M50, *Water Resources Planning*, #30050
M51, *Air Valves: Air-Release, Air/Vacuum and Combination*, #30051
M52, *Water Conservation Programs—A Planning Manual*, #30052
M53, *Microfiltration and Ultrafiltration Membranes for Drinking Water*, #30053
M54, *Developing Rates for Small Systems*, #30054
M55, *PE Pipe—Design and Installation*, #30055
M56, *Nitrification Prevention and Control in Drinking Water*, #30056
M57, *Algae: Source to Treatment*, #30057
M58, *Internal Corrosion Control in Water Distribution Systems*, #30058
M60, *Drought Preparedness and Response*, #30060
M61, *Desalination of Seawater*, #30061
M62, *Membrane Applications for Water Reuse*, #30062
M63, *Aquifer Storage and Recovery*, #30063

M65, *On-Site Generation of Hypochlorite,* #30065

M66, *Cylinder and Vane Actuators and Controls—Design and Installation,* #30066

M68, *Water Quality in Distribution Systems,* #30068

M69, *Inland Desalination and Concentrate Management,* #30069

M77, *Condition Assessment of Water Mains,* #30077

www.ingramcontent.com/pod-product-compliance
Lightning Source LLC
LaVergne TN
LVHW080044210926
848481LV00011B/3276
9781625763310